Study Guide

Statistics for the Behavioral Sciences

EIGHTH EDITION

Frederick J Gravetter
State University of New York, Brockport

Larry B. Wallnau

Prepared by

Frederick J Gravetter
State University of New York, Brockport

Australia • Brazil • Japan • Korea • Mexico • Singapore • Spain • United Kingdom • United States

© 2010 Wadsworth, Cengage Learning

ALL RIGHTS RESERVED. No part of this work covered by the copyright herein may be reproduced, transmitted, stored, or used in any form or by any means graphic, electronic, or mechanical, including but not limited to photocopying, recording, scanning, digitizing, taping, Web distribution, information networks, or information storage and retrieval systems, except as permitted under Section 107 or 108 of the 1976 United States Copyright Act, without the prior written permission of the publisher.

For product information and technology assistance, contact us at
**Cengage Learning Customer & Sales Support,
1-800-354-9706**

For permission to use material from this text or product, submit all requests online at **www.cengage.com/permissions**
Further permissions questions can be emailed to
permissionrequest@cengage.com

ISBN-13: 978-0-495-60296-5
ISBN-10: 0-495-60296-5

Wadsworth
10 Davis Drive
Belmont, CA 94002-3098
USA

Cengage Learning is a leading provider of customized learning solutions with office locations around the globe, including Singapore, the United Kingdom, Australia, Mexico, Brazil, and Japan. Locate your local office at: **www.cengage.com/international**

Cengage Learning products are represented in Canada by Nelson Education, Ltd.

To learn more about Wadsworth, visit
www.cengage.com/wadsworth

Purchase any of our products at your local college store or at our preferred online store
www.ichapters.com

Printed in the United States of America
1 2 3 4 5 6 7 12 11 10 09 08

CONTENTS

How To Use This Study Guide v

Chapter 1: Introduction to Statistics 1

Chapter 2: Frequency Distributions 13

Chapter 3: Central Tendency 28

Chapter 4: Variability 41

Chapter 5: z-scores 51

Chapter 6: Probability 62

Chapter 7: The Distribution of Sample Means 73

Chapter 8: Introduction to Hypothesis Testing 85

Chapter 9: Introduction to the t Statistic 98

Chapter 10: The t Test for Two Independent Samples 110

Chapter 11: The t Test for Two Related Samples 122

Chapter 12: Estimation 136

Chapter 13: Introduction to Analysis of Variance 147

Chapter 14: Repeated-Measures Analysis of Variance 161

Chapter 15: Two-Factor Analysis of Variance 177

Chapter 16: Correlation 194

Chapter 17: Introduction to Regression 208

Chapter 18: Hypothesis Tests with Chi-Square 223

Chapter 19: The Binomial Test 236

Chapter 20: Statistical Tests for Ordinal Data 245

How to Use
This Study Guide

As the title indicates, the purpose of a Study Guide is to provide you with a guide for studying. In general, studying involves revisiting and reviewing material that you have already seen in class and in the textbook. Notice that we have used the word "review," as in "a second look." The textbook is written to provide an introduction to statistical concepts, and includes all the explanations and examples necessary to present new ideas that students are seeing for the first time. This *Study Guide*, on the other hand, is written to provide a second look at material that is already somewhat familiar. That is, the Study Guide is intended as a "review" that condenses, summarizes, and highlights the textbook. Thus, the Study Guide is a supplement to the textbook, not a replacement for the textbook, and we strongly suggest that you work in the Study Guide only after you have read and studied the corresponding section in the textbook.

Each chapter in the Study Guide corresponds to a chapter in the textbook, and each Study Guide chapter is divided into seven major sections. The following paragraphs provide an overview of these sections and some suggestions for how they should be used.

Chapter Summary We begin with a big-picture overview of the chapter contents. The intent is to provide a relatively concise summary of the general content of the chapter and to identify most of the major points. This is the kind of thing that your mother wanted to know when she asked, "What did you learn in school today?" If you understand the big picture, then all the details, terminology, and formulas will be much easier to learn. Keep in mind, however, that the summary does not include all of the individual topics that are covered in the chapter. You will need to read the complete chapter and work some of the problems before you will pick up all the details.

Learning Objectives. Next we provide a list of goals that should be achieved upon completion of studying a text chapter. These objectives usually are task-oriented, pointing to specific things you should know how to do or how to explain. If you lack confidence in meeting any of these objectives, you should return to the appropriate section(s) of the text for further study.

New Terms And Concepts This section consists of a list of important terms and concepts that appeared in the text. You should identify or define these terms, and explain how closely-related terms are interrelated. Return to the text to check your answers.

New Formulas A list of formulas that were introduced in the text is provided. For each formula, you should test yourself in the following way:
1) Identify each symbol and term in the formula.
2) Make a list of the computational steps that are indicated by the formula.
3) Describe when each formula is used and explain why it is used or what it is computing.
4) Most importantly, you should realize that each formula is simply a concise, mathematical expression of a concept or procedure. If you can explain the concept or procedure in words, then it is usually very easy to translate the words into mathematical symbols and recreate the formula. Thus, we encourage you to understand concepts rather than memorize formulas.

Step-By-Step This section presents a typical problem (or problems) from the chapter and provides a demonstration of the step-by-step procedures for solving the problem.

Hints and Cautions. In this section, we provide advice on the typical mistakes students make, and the difficulties they commonly have with the chapter material.

Self-Test. In this section we present a series of questions and problems that provide a general review of the chapter. As much as possible, the problems consist of simple sets of numbers so that you can solve them with minimal calculation. Work through the Self-Test carefully, and remember that your mistakes will tell you what areas need additional study. Incidentally, the **Answers to the Self-Test** are given at the end of each Study Guide chapter. For problems with numerical answers, do not fret about slight differences between your answer and ours. A little rounding error is to be expected.

STUDY HINTS

It seems appropriate for a study guide to have a few suggestions to help you study. The following are some hints that have proved useful for our own students.

1. You will learn (and remember) much more if you study for short periods several times a week, rather than concentrating all your studying into one long session. For example, it is far more effective to study for one-half hour every night than to spend a single three and one-half hour session once a week.

2. Do some work before class. Read the appropriate sections in the textbook before your instructor presents the material in class. Although you may not completely understand what you read, you will have a general idea of the topic which will make the lecture easier to follow. Also, you can identify items that are particularly confusing and then be sure that these items are clarified in class.

3. Pay attention and think during class. Although this may sound like obvious advice, many students spend their class time frantically taking notes rather than listening and understanding what is being said. For example, it usually is not necessary to copy down every sample problem that your instructor works in class. There are plenty of example problems in your textbook and in this study guide - you probably do not need one more in your notebook. Just put down your pencil and pay attention. Be sure that you understand what is being done, and try to anticipate the next step in the problem.

4. Test yourself regularly. Do not wait until the end of the chapter or the end of the week to check your knowledge. After each lecture, work some of the end-of-chapter problems, do the learning checks, and be sure you can define key terms. If you are having trouble, get your questions answered immediately (re-read the text, go to your instructor, ask questions in class). Do not let yourself fall behind.

5. Don't kid yourself. Many students sit in class watching the instructor solve problems, and think to themselves, "This looks easy, I understand it." Do you really? Can you do the problem all by yourself? Many students use the examples in the textbook as a guide for working assigned problems. They begin the problem, get stuck, then check the example to see what to do next. A minute later they are stuck again, so they take another peek at the example in the text. Eventually, the problem is finished and the students think that they understand how to solve problems. Although there is nothing wrong with using examples as models for solving problems, you should try working a problem with your book closed to determine whether you can complete it on your own.

PREPARING FOR EXAMS

1. Perhaps the best way to get ready for an exam is to make up your own exam. This is particularly effective if you have a friend in the class so you can both make up exams and then exchange them. Constructing your own exam forces you to identify the important points in the material, and it makes you think about how exam questions might be phrased. It is very satisfying to open a statistics exam and find some questions that you wrote yourself the day before.

2. Many students suffer from "exam anxiety" which causes them to freeze up and forget everything during exams. One way to help avoid the problem is for you to take charge of your own time during an exam.
    a. Don't spend a lot of time working on one problem that you really don't understand (especially if it is a 1-point true/false question). Just move on to the rest of the exam - you can come back later, if you have time.

b. Remember that you do not have to finish the exam questions in the order they are presented. When you get your exam, go immediately to the problems you understand best. This will build some confidence and make you better prepared for the remainder of the exam.

Probably the best way to reduce exam anxiety is to practice taking exams. Make up your own exam (select problems from the book and study guide) or have a friend make up an exam. Then try to duplicate the general conditions of an exam. If you are not allowed to use your book during exams, then put it away during practice. Give yourself a time limit. If you have an old alarm clock, set it in front of you so that you can hear the ticking and watch the time slip away. You might even try sitting in front of a mirror so that every time you look up there is someone watching you.

Finally, remember that very few miracles happen during exams. The work on your exam is usually a good reflection of your studying and understanding. Most students walk into an exam with a very good idea of how well they will do. Be honest with yourself. If you are well prepared, you will do well on the exam, and there is no reason to panic. If you are not prepared, then exam anxiety probably is not your problem.

CHAPTER 1

INTRODUCTION TO STATISTICS

CHAPTER SUMMARY

The goals of Chapter 1 are:
1. To introduce the basic terminology that will be used in statistics.
2. To explain how statistical techniques fit into the general process of scientific research.
3. To introduce some of the notation that will be used throughout the rest of the book.

Terminology

A **variable** is a characteristic or condition that can change or take on different values. Most research begins with a general question about the relationship between two variables for a specific group of individuals. The entire group of individuals is called the **population**. For example, a researcher may be interested in the relation between class size (variable 1) and academic performance (variable 2) for the population of third-grade children. Usually populations are so large that a researcher cannot examine the entire group. Therefore, a **sample** is selected to represent the population in a research study. The goal is to use the results obtained from the sample to help answer questions about the population.

Variables can be classified as discrete or continuous. **Discrete variables** (such as class size) consist of indivisible categories, and **continuous variables** (such as time or weight) are infinitely divisible into whatever units a researcher may choose. For example, time can be measured to the nearest minute, second, half-second, etc.) To define the units for a continuous variable, a researcher must use **real limits** which are boundaries located exactly half-way between adjacent categories.

To establish relationships between variables, researchers must observe the variables and record their observations. This requires that the variables be **measured**. The process of measuring a variable requires a set of categories called a **scale of measurement** and a process that classifies each individual into one category. Four types of measurement scales are as follows:
  a. A **nominal scale** is an unordered set of categories identified only by name. Nominal measurements only permit you to determine whether two individuals are the same or different.
  b. An **ordinal scale** is an ordered set of categories. Ordinal measurements tell you the direction of difference between two individuals.
  c. An **interval scale** is an ordered series of equal-sized categories. Interval measurements identify the direction and magnitude of a difference. The zero point is located arbitrarily on an interval scale.

d. A **ratio scale** is an interval scale where a value of zero indicates none of the variable. Ratio measurements identify the direction and magnitude of differences and allow ratio comparisons of measurements.

Two broad categories of research studies are experiments and correlational studies. The goal of a **correlational** study is to determine whether there is a relationship between two variables and to describe the relationship. A **correlational** study simply observes the two variables as they exist naturally. The goal of an **experiment** is to demonstrate a cause-and-effect relationship between two variables; that is, to show that changing the value of one variable causes changes to occur in a second variable. In an **experiment**, one variable is manipulated to create treatment conditions. A second variable is observed and measured to obtain scores for a group of individuals in each of the treatment conditions. The measurements are then compared to see if there are differences between treatment conditions. All other variables are controlled to prevent them from influencing the results. In an experiment, the manipulated variable is called the **independent variable** and the observed variable is the **dependent variable**.

Other types of research studies, know as **non-experimental** or **quasi-experimental**, are similar to experiments because they also compare groups of scores. However, these studies do not use a manipulated variable to differentiate the groups. Instead, the variable that differentiates the groups is usually a pre-existing participant variable (such as male/female) or a time variable (such as before/after). Because these studies do not use the manipulation and control of true experiments, they cannot demonstrate cause and effect relationships. As a result, they are similar to correlational research because they simply demonstrate and describe relationships.

Statistics in Science

The measurements obtained in a research study are called the **data**. The goal of statistics is to help researchers organize and interpret the data. Statistical techniques are classified into two broad categories: descriptive and inferential. **Descriptive statistics** are methods for organizing and summarizing data. For example, tables or graphs are used to organize data, and descriptive values such as the average score are used to summarize data. A descriptive value for a population is called a **parameter** and a descriptive value for a sample is called a **statistic**. **Inferential statistics** are methods for using sample data to make general conclusions (inferences) about populations. Because a sample is typically only a part of the whole population, sample data provide only limited information about the population. As a result, sample statistics are generally imperfect representatives of the corresponding population parameters. The discrepancy between a sample statistic and its population parameter is called **sampling error**. The concept of sampling error is demonstrated in Figure 1.2 in the textbook. Defining and measuring sampling error is a large part of inferential statistics.

Notation

The individual measurements or scores obtained for a research participant will be identified by the letter X (or X and Y if there are multiple scores for each individual). The number of scores in a data set will be identified by N for a population or n for a sample.

Summing a set of values is a common operation in statistics and has its own notation. The Greek letter sigma, Σ, will be used to stand for "the sum of." For example, ΣX identifies the sum of the scores. To use and interpret summation notation, you must follow the basic **order of operations** required for all mathematical calculation.

1. All calculations within parentheses are done first.
2. Squaring or raising to other exponents is done second.
3. Multiplying, and dividing are done third, and should be completed in order from left to right.
4. Summation with the Σ notation is done next.
5. Any additional adding and subtracting is done last and should be completed in order from left to right.

For example, to compute $\Sigma X + 3$, you sum the X values, then add 3.
To compute $\Sigma(X + 3)^2$, you add 3 to each X (inside parentheses), then square the resulting values, then sum the squared numbers.

LEARNING OBJECTIVES

1. You should be familiar with the terminology and special notation of statistical methods.

2. You should understand the purpose of statistics: When, how, and why they are used.

3. You should understand summation notation and be able to use this notation to represent mathematical operations and to compute specified sums.

NEW TERMS AND CONCEPTS

The following terms were introduced in Chapter 1. You should be able to define or describe each term and, where appropriate, describe how each term is related to other terms in the list.

Population	The entire group of individuals that a researcher wishes to study.
Sample	A group selected from a population to participate in a research study.
Statistic	A Characteristic that describes a sample.
Parameter	A characteristic that describes a population.
Sampling Error	The discrepancy between a statistic and a parameter.
Descriptive Statistics	Techniques that organize and summarize a set of data
Inferential statistics	Techniques that use sample data to draw general conclusions about populations.

Variable	A characteristic that can change or take on different values.
Constant	A characteristic that does not change.
Raw score	An original, unaltered measurement.
Dependent variable	In an experiment, the variable that is observed for changes. (the scores)
Independent variable	In an experiment, the variable that is manipulated by the researcher. (the treatment conditions)
Correlational method	A research method that simply observes two existing variables to determine the nature of their relationship.
Experimental method	A research method that manipulates one variable, observes a second variable for changes, and controls all other variables. The goal is to establish a cause-and-effect relationship.
Control condition	A condition where the treatment is not administered.
Experimental condition	A condition where the treatment is administered.
Hypothetical construct	A characteristic or mechanism that is assumed to exist but cannot be observed or measured directly.
Operational definition	A procedure for measuring and defining a construct.
Nominal scale	A measurement scale where the categories are differentiated only by qualitative names.
Ordinal scale	A measurement scale consisting of a series of ordered categories.
Interval scale	An ordinal scale where all the categories are intervals with exactly the same width.
Ratio scale	An interval scale where a value of zero corresponds to none.
Discrete variable	A variable that exists in indivisible units.
Continuous variable	A variable that can be divided into smaller units without limit.
Real limits	The boundaries separating the intervals that define the scores for a continuous variable.

Σ	Summation sign - the sum of.
N	The number of scores in a population.
n	The number of scores in a sample.
X	A score.

STEP BY STEP

Summation Notation: In statistical calculations you constantly will be required to add a set of values to find a specific total. We will use algebraic expressions to represent the values being added (for example, X = score), and we will use the Greek letter sigma (Σ) to signify the process of summation. Occasionally, you simply will be adding a set of scores, ΣX. More often, you will be doing some initial computation and then adding the results. For example, we will routinely need to square each score and then find the sum of the squared values, ΣX^2. The following step-by-step process should help you understand summation notation and use it correctly to find appropriate totals.

Step 1: A mathematical expression that includes summation notation (Σ) can be viewed as a set of step-by-step instructions. Your job is to identify the different steps and then perform them in the proper order. To accomplish this, you must follow the order of operations presented earlier (page 3).

Step 2: Set up a computational table listing the original X values in the first column. As you perform each step in the order of operations, you will create a new column showing the results of that step.

Suppose your task is to find $\Sigma(X + 3)$. The first step according to these instructions is to add three points to each X values (inside parentheses). Start a new column headed with (X + 3), and list each value you obtain as you add three points to the original X scores.

X	(X + 3)
4	7
8	11
2	5
6	9

Note: Occasionally you will need more than one column to get to the final term you want. To compute $\Sigma(X + 3)^2$, for example, begin with the X column, then add a column of (X + 3) values, and then add a third column that squares each of the (X + 3) values. Thus, each column represents a step in the computations.

X	(X + 3)	(X + 3)²
4	7	49
8	11	121
2	5	25
6	9	81

Step 3: When you reach the point where the next step in the order of operations is the summation, simply add all the values in the appropriate column.

Using the same numbers that we used in Step 2, you find $\Sigma(X + 3)$ by simply adding the values in the (X + 3) column.

$$\Sigma(X + 3) = 7 + 11 + 5 + 9 = 32$$

To find $\Sigma(X + 3)^2$ you add the values in the (X + 3)² column.

$$\Sigma(X + 3)^2 = 49 + 121 + 25 + 81 = 276.$$

HINTS AND CAUTIONS

1. Many students confuse the independent variable and the dependent variable in an experiment. It may help you to differentiate these terms if you visualize the results (the data) from a typical experiment.

	Treatment 1	Treatment 2
Independent Variable Manipulated to create → Treatment Conditions	12	24
	10	21
Dependent Variable	14	29
The scores measured →	16	23
In each treatment	11	22

The goal is to determine whether the scores depend on the treatments. In this example, the scores in Treatment 2 appear to be higher than the scores in Treatment 1.

2. There are three specific sums that are used repeatedly in statistics calculations. You should know the notation and computations for each of the following:
   a. ΣX^2 First square each score, then add the squared values.
   b. $(\Sigma X)^2$ First sum the scores, then square the total.
   c. $\Sigma(X - C)^2$ First subtract the constant C from each score, then square each of the resulting values. Finally, add the squared numbers.

SELF TEST

True/False Questions

1. A researcher is interested in the average income for registered voters in the United States. The entire group of registered voters is an example of a population.

2. A researcher interested in vocabulary development obtains a sample of 3-year-old children to participate in a research study. The average score for the group of 20 is an example of a parameter.

3. Statistical techniques that use sample data to answer questions about populations are known as descriptive statistics.

4. The goal of an experiment is to demonstrate the existence of a cause-and-effect relationship between two variables.

5. A correlational study typically uses only one group of participants but measures two different variables (two scores) for each individual.

6. A researcher is using an experiment to determine whether sugar consumption has any effect on the activity level of preschool children. For this study, the dependent variable is the activity level of the children.

7. Classifying people into two groups on the basis of gender is an example of measurement on an ordinal scale.

8. A discrete variable can only be measured in indivisible categories.

9. For the following scores, $\Sigma(X-1) = 10$. Scores: 1, 3, 7

10. For the following scores, $\Sigma X^2 = 35$. Scores: 1, 3, 5

Multiple-Choice Questions

1. Although research questions usually concern a _____, the actual research is typically conducted with a _____.
    a. sample, statistic
    b. population, parameter
    c. sample, population
    d. population, sample

INTRODUCTION TO STATISTICS

2. A researcher is interested in the eating behavior of rats and selects a group of 25 rats to be tested in a research study. The average score for the group of 25 rats is an example of a _____.
   a. sample
   b. statistic
   c. population
   d. parameter

3. A researcher conducts an experiment to determine whether moderate doses of St. Johns Wort have any effect of memory for college students. For this study, what is the independent variable?
   a. the amount of St. Johns Wort given to each participant
   b. the memory score for each participant
   c. the group of college students
   d. cannot answer without more information

4. One group of new freshmen is given a study-skills training course during the first week of college and a second group does not receive the course. At the end of the semester, the grade point average is recorded for each student. For this study, what is the dependent variable?
   a. receiving or not receiving the course
   b. the grade point average for each student
   c. the group that received the training course
   d. the group that did not receive the course

5. A researcher wants to examine the relationship between sleeping habits and grade point average for college students. A group of 100 students is selected and each student is asked to report the average number of hours he/she sleeps each night and his/her grade point average. What research method is being used in this study?
   a. experimental method
   b. correlational method
   c. non-experimental method comparing two groups
   d. none of the above

6. In a study evaluating the effectiveness of a new medication designed to control high blood pressure, one sample of individuals is given the medicine and a second sample is given an inactive placebo. Blood pressure is measured for each individual. For this study, what is the dependent variable?
   a. the medication
   b. the placebo
   c. blood pressure
   d. the individuals given the medication

7. A typical experimental research study
   a. measures 1 variable for each participant and uses only 1 group
   b. measures 2 variables for each participant and uses only 1 group
   c. measures 1 variable for each participant and compares 2 groups of scores
   d. measures 2 variables for each participant and compares 2 groups of scores

8. To determine the size of the difference between two measurements, a researcher must use a(n) _____ scale of measurement.
    a. nominal
    b. nominal or ordinal
    c. ratio
    d. interval or ratio

9. Determining the class standing (1st, 2nd, and so on) for the graduating seniors at a high school would involve measurement on a(n) _____ scale of measurement.
    a. nominal
    b. ordinal
    c. interval
    d. ratio

10. Which of the following is an example of a discrete variable?
    a. height
    b. reaction time
    c. number of brothers and/or sisters
    d. age

11. Real limits are an important consideration when measuring a _____ variable.
    a. ratio scale
    b. ratio or interval scale
    c. discrete
    d. continuous

12. What is the final step to be performed when computing $\Sigma(X - 2)^2$?
    a. square each value
    b. subtract 2 points from each score
    c. sum the squared values
    d. subtract 2^2 from each X^2 value

13. What is the value of $(\Sigma X)^2$ for the following scores?   Scores: 1, 5, 2
    a. 10
    b. 16
    c. 30
    d. 64

14. What is the value of $\Sigma(X + 1)$ for the following scores?   Scores: 2, 0, 4, 2
    a. 9
    b. 11
    c. 12
    d. 36

15. You are instructed to add 1 point to each score, then square the resulting values, and then find the sum of the squared numbers. In summation notation, this set of operations would be expressed as
    a. $(\Sigma X + 1)^2$
    b. $\Sigma(X + 1)^2$
    c. $\Sigma(X^2 + 1)$
    d. $\Sigma X^2 + 1$

Other Questions

1. Describe the relationships between a *sample*, a *population*, a *statistic* and a *parameter*.

2. What are the basic characteristics of an experiment that differentiate this method from other types of research?

3. Compute each value requested for the following set of scores.

X	
1	ΣX
3	ΣX^2
5	$(\Sigma X)^2$
2	N

4. Compute each value requested for the following set of scores.

X	
0	$\Sigma X + 1$
6	$\Sigma(X + 1)$
2	$\Sigma(X + 1)^2$
3	

5. Use summation notation to express each of the following calculations.
    a. Add 3 points to each score, then find the sum of the resulting values.
    b. Find the sum of the scores, then add 10 points to the total.
    c. Subtract 1 point from each score, then square each of the resulting values. Next, find the sum of the squared numbers. Finally, add 5 points to this sum.

ANSWERS TO SELF TEST

True/False Answers

1. True.
2. False. The average for a sample is a statistic
3. False. They are inferential statistics.
4. True.
5. True.
6. True.
7. False. Gender is measured on a nominal scale.
8. True.
9. False. Subtract 1 point from each score, then add.
10. True.

Multiple-Choice Answers

1. d Populations are usually too large to study so researchers use samples.
2. b A characteristic that describes a sample is called a statistic.
3. a. The independent variable is manipulated by the researcher.
4. b The dependent variable is the score measured for each individual.
5. b The correlational method simply measures two variables for each participant.
6. c The dependent variable is the score measured for each individual.
7. c A typical experiment has one independent variable (defining the groups) and one dependent variable (the scores).
8. d An interval or ratio scale is needed to measure distance.
9. b An ordinal scale consists of ranks.
10. c Only whole numbers are possible with no intermediate values.
11. d Continuous variables are divided into intervals with real limits.
12. c Parentheses and squaring both have priority over summation.
13. d Sum the scores, then square the sum.
14. c Add 1 point to each score, then sum.
15. b Adding 1 point is done in parentheses, then square, then add.

Other Answers

1. A population is the entire group of individuals that you are interested in studying. A sample is a group selected from the population to participate in the research study. A parameter is a characteristic, usually a numerical value, of a population. A statistic is a characteristic of a sample.

2. The goal of an experiment is to establish a cause-and-effect relationship between two variables. To accomplish this goal, an experiment has two distinguishing features. First, the researcher manipulates one variable (the independent variable) to create the different treatment conditions that will be compared. The dependent variable is the score that is measured inside each treatment condition. The second feature of an experiment is that all other variables are controlled so that they cannot influence the independent variable or the dependent variable.

3. $\Sigma X = 11$, $\Sigma X^2 = 39$, $(\Sigma X)^2 = 121$, $N = 4$

4. $\Sigma X + 1 = 11 + 1 = 12$
  $\Sigma(X + 1) = 1 + 7 + 3 + 4 = 15$
  $\Sigma(X + 1)^2 = 1 + 49 + 9 + 16 = 75$

5. a. $\Sigma(X + 3)$
   b. $\Sigma X + 10$
   c. $\Sigma(X - 1)^2 + 5$

CHAPTER 2

FRENQUENCY DISTRIBUTIONS

CHAPTER SUMMARY

The goals of Chapter 2 are:
1. To introduce the concept of a frequency distribution as a descriptive statistical technique.
2. To describe how a frequency distribution can be presented in a regular or grouped table.
3. To describe how a frequency distribution can be presented in a graph.
4. To introduce the concept of percentiles and percentile ranks, and the mathematical process of interpolation.
5. To introduce stem-and-leaf displays.

Frequency Distributions

After collecting data, the first task for a researcher is to organize and simplify the data so that it is possible to get a general overview of the results. This is the goal of descriptive statistical techniques. One method for simplifying and organizing data is to construct a frequency distribution. A **frequency distribution** is an organized tabulation showing exactly how many individuals are located in each category on the scale of measurement. A frequency distribution presents an organized picture of the entire set of scores, and it shows where each individual is located relative to others in the distribution.

Frequency Distribution Tables

A **frequency distribution table** consists of at least two columns - one listing categories on the scale of measurement (X) and another for frequency (f). In the X column, values are listed from the highest to lowest, without skipping any. For the frequency column, tallies are determined for each value (how often each X value occurs in the data set). These tallies are the frequencies for each X value. The sum of the frequencies should equal N. A third column can be used for the proportion (p) for each category: $p = f/N$. The sum of the p column should equal 1.00. Finally, a fourth column can display the percentage of the distribution corresponding to each X value. The percentage is found by multiplying p by 100. The sum of the percentage column is 100%.

When a frequency distribution table lists all of the individual categories (X values) it is called a **regular frequency distribution**. Sometimes, however, a set of scores covers a wide range of values. In these situations, a list of all the X values would be quite long - too long to be a "simple" presentation of the data. To remedy this situation, a **grouped frequency distribution** table is used. In a grouped table, the X column lists groups of scores, called **class intervals**, rather than individual values. These intervals all have the same width, usually a simple number such as 2, 5, 10, and so on. Each interval begins with a value that is a multiple of the interval width. The interval width is selected so that the table will have approximately ten intervals.

Frequency Distributions Graphs

In a **frequency distribution graph**, the score categories (X values) are listed on the X axis and the frequencies are listed on the Y axis. When the score categories consist of numerical scores from an interval or ratio scale, the graph should be either a histogram or a polygon. In a **histogram**, a bar is centered above each score (or class interval) so that the height of the bar corresponds to the frequency and the width extends to the real limits, so that adjacent bars touch. In a **polygon**, a dot is centered above each score so that the height of the dot corresponds to the frequency. The dots are then connected by straight lines. An additional line is drawn at each end to bring the graph back to a zero frequency.

When the score categories (X values) are measurements from a nominal or an ordinal scale, the graph should be a bar graph. A **bar graph** is just like a histogram except that gaps or spaces are left between adjacent bars.

If it is possible to count the number of scores in a population, then the population distribution can be presented in a histogram, polygon, or bar graph. However, many populations are so large that it is impossible to know the exact number of individuals (frequency) for any specific category. In these situations, population distributions can be shown using **relative frequency** instead of the absolute number of individuals for each category. If the scores in the population are measured on an interval or ratio scale, it is customary to present the distribution as a **smooth curve** rather than a jagged histogram or polygon. The smooth curve emphasizes the fact that the distribution is not showing the exact frequency for each category.

Frequency distribution graphs are useful because they show the entire set of scores. At a glance, you can determine the highest score, the lowest score, and where the scores are centered. The graph also shows whether the scores are clustered together or scattered over a wide range. Finally, a graph shows the **shape** of the distribution. A distribution is **symmetrical** if the left side of the graph is (roughly) a mirror image of the right side. One example of a symmetrical distribution is the bell-shaped normal distribution. On the other hand, distributions are **skewed** when scores pile up on one side of the distribution, leaving a "tail" of a few extreme values on the other side. In a **positively skewed** distribution, the scores tend to pile up on the left side of the distribution with the tail tapering off to the right. In a **negatively skewed** distribution, the scores tend to pile up on the right side and the tail points to the left.

Percentiles, Percentile Ranks, and Interpolation

The relative location of individual scores within a distribution can be described by percentiles and percentile ranks. The **percentile rank** for a particular X value is the percentage of individuals with scores equal to or less than that X value. When an X value is described by its rank, it is called a **percentile**. For example, if 80% of the scores in a distribution are lower than your score (and 20% are higher), then you score has a percentile rank of 80% and your score is called the 80th percentile.

To find percentiles and percentile ranks, two new columns are placed in the frequency distribution table: One is for cumulative frequency (cf) and the other is for cumulative percentage (c%). Each cumulative percentage identifies the percentile rank for the upper real limit of the corresponding score or class interval. When scores or percentages do not correspond to upper real limits or cumulative percentages, you must use interpolation to determine the corresponding ranks and percentiles. **Interpolation** is a mathematical process based on the assumption that the scores and the percentages change in a regular, linear fashion as you move through an interval from one end to the other.

Stem-and-Leaf Displays

A **stem-and-leaf display** provides a very efficient method for obtaining and displaying a frequency distribution. Each score is divided into a **stem** consisting of the first digit or digits, and a **leaf** consisting of the final digit. For example, X = 36 has a stem of 3 and a leaf of 6. You then list the complete set of stems for a set of scores. Finally, you go through the list of scores, one at a time, and write the leaf for each score beside its stem. The resulting display provides an organized picture of the entire distribution. The number of leafs beside each stem corresponds to the frequency, and the individual leafs identify the individual scores.

LEARNING OBJECTIVES

1. Know how to organize data into regular or grouped frequency distribution tables.

2. Be able to construct graphs, including bar graphs, histograms, and polygons.

3. Be able to describe the shape of a distribution portrayed in a frequency distribution graph.

4. Be able to describe locations within a distribution using percentiles or percentile ranks, and be able to compute percentiles and percentile ranks using interpolation when necessary.

NEW TERMS AND CONCEPTS

The following terms were introduced in this chapter. You should be able to define or describe each term and, where appropriate, describe how each term is related to other terms in the list.

Frequency distribution	A tabulation of the number of individuals in each category on the scale of measurement.
Grouped frequency distribution	A frequency distribution where scores are grouped into intervals rather than listed as individual values.
Class interval	A group of scores in a grouped frequency distribution.
Upper real limit	The boundary that separates an interval from the next higher interval.
Lower real limit	The boundary that separates an interval from the next lower interval.

Apparent limits	The score values that appear as the lowest score and the highest score in an interval.
Histogram	A graph showing a bar above each score or interval so that the height of the bar corresponds to the frequency and width extends to the real limits.
Polygon	A graph consisting of a line that connects a series of dots. A dot is placed above each score or interval so that the height of the dot corresponds to the frequency.
Bar graph	A graph showing a bar above each score or interval so that the height of the bar corresponds to the frequency. A space is left between adjacent bars.
Relative frequency	The proportion of the total distribution rather than the absolute frequency. Used for population distributions for which the absolute number of individuals is not known for each category.
Symmetrical distribution	A distribution where the left-hand side is a mirror image of the right-hand side.
Positively skewed distribution	A distribution where the scores pile up on the left side and taper off to the right.
Negatively skewed distribution	A distribution where the scores pile up on the right side and taper off to the left.
Tail of a distribution	A section on either side of a distribution where the frequency tapers down toward zero as the X values become more extreme.
Percentile rank	The percentile rank of a specific score is the percentage of individuals with scores at or below the specific score.
Percentile	A score that is described in terms of its percentile rank.
Interpolation	A mathematical procedure for estimating a value that is located between two known values.
Stem-and-leaf display	A technique for presenting a complete distribution of scores in an organized display similar to a grouped frequency distribution table.

NEW FORMULAS

$$\text{proportion} = p = f/N \qquad \text{percentage} = p(100) = (f/N)(100)$$

STEP BY STEP

<u>Constructing a Frequency Distribution Table</u>: The goal of a frequency distribution table is to take an entire set of scores and simplify and organize them into a form that allows a researcher to see at a glance the entire distribution. Suppose, for example, that an instructor gave a personality questionnaire measuring self-esteem to an entire class of psychology students. The questionnaire classifies each individual into one of five categories indicating different levels of self esteem: 1 = high self-esteem and 5 = low self-esteem.
The results for the class are as follows:

    4, 4, 3, 3, 5, 4, 2, 1, 1, 3
    4, 4, 5, 2, 3, 3, 4, 3, 3, 2
    1, 4, 5, 3, 4, 4, 5, 2, 4, 1
    3, 3, 2, 2, 4, 5, 1, 5, 3, 4

Step 1: In the first column of the table, list the categories that make up the scale of measurement starting with the highest score at the top and listing every possible X value down to the lowest score. The column heading should be "X" to indicate that this is the scale of X values.

X
5
4
3
2
1

Step 2: In a second column, headed by "f" for frequency, list the number of individuals who have each score. For example, six people have scores of X = 5, so you place a 6 in the f column beside the X value 5. Continue for each score (category) on the scale of measurement.

X	f
5	6
4	12
3	11
2	6
1	5

The result is a basic frequency distribution table. The table can be expanded by adding columns for proportion or percentage.

<u>Retrieving Scores from a Frequency Distribution Table</u>: Although a frequency distribution table provides a concise overview of an entire set of data, it is a condensed version of the actual data and for some students the table can obscure the details of the individual scores.

For example, in the table that we have just constructed, the 40 individual scores have been condensed into a table that shows only 10 numerical values (five X's and five frequencies). For some purposes, it is easier to transform a frequency distribution table back into a complete set of scores before you begin any statistical calculations with the data. We will use the table we have already constructed to demonstrate the process of recovering individual scores from a frequency distribution table.

Step 1: Find the number of individual scores (N): The frequency column (f) of the table shows the number of individuals located in each category on the scale of measurement. For this example, six individuals had scores of X = 5, twelve individuals had scores of X = 4, and so on. To determine the total number of individuals in the group, you simply add the frequencies.

$N = \Sigma f = 6 + 12 + 11 + 6 + 5 = 40$

Step 2: List the complete set of individual scores: Again, the table shows that six individual had scores of X = 5, twelve individuals had X = 4, and so on. These scores can be listed individually as follows:

5, 5, 5, 5, 5, 5   six fives
4, 4, 4, 4, 4, 4, 4, 4, 4, 4, 4, 4   twelve fours
3, 3, 3, 3, 3, 3, 3, 3, 3, 3, 3   eleven threes
2, 2, 2, 2, 2, 2   six twos
1, 1, 1, 1, 1   five ones

With the complete set of N = 40 scores listed in this way, it is easy to perform computations based on individual scores. For example, to find ΣX you would add the 40 values ($\Sigma X = 128$).

Interpolation: Because it is impossible to report an infinite number of data points, nearly all tables and graphs show only a limited number of selected values. However, researchers often want to examine data points that fall between the reported values. The process of interpolation provides a method for estimating intermediate values.

In this chapter we used interpolation to find percentiles and percentile ranks that cannot be read directly from a frequency distribution table. The following example will be used to demonstrate this process.

The problem is to find the 50th percentile for the distribution shown in the following table. Because 50% is not one of the cumulative percentages listed in the table (it is between 20% and 65%), we must use interpolation.

X	f	cf	c%
25-29	1	20	100%
20-24	1	19	95%
15-19	5	18	90%
10-14	9	13	65%
5-9	4	4	20%

Step 1: Identify the interval that contains the intermediate value you want. In this example we are looking for the score that has a rank of 50%. The intermediate value, 50%, is located between the reported values of 20% and 65%.

Step 2: Draw a sketch showing the interval you identified in Step 1. Show the two end points of the interval and identify the location of the intermediate value.

    c%
   65%
     ---50%
   20%

Step 3: Expand your sketch by placing the second scale beside the one you have drawn. Then find the end-points of the interval on the second scale. Remember, each cumulative percentage value is associated with the upper real limit of the score interval.

     X   c%
   14.5   65%
?----   ---50%

    9.5   20%

Step 4: Look at your sketch and make a common-sense estimate of the final answer. In this example, the value we want is between 14.5 and 9.5. It appears to be closer to 14.5, probably around X = 13. (This kind of preliminary estimate can save you from making a careless mistake later.)

Step 5: Find the precise location of the intermediate value within the interval. This step requires that you compute a fraction,

$$\text{fraction} = \frac{\text{distance from the top of the interval}}{\text{interval width}}$$

For this example, the intermediate value of 50% is located 15 points from the top of the interval and the interval width is 45 points (from 65% to 20%). Thus,

fraction = 15/45 = 1/3 = 0.33

The position we want is located 1/3rd of the way down from the top of the interval.

Step 6: Apply the fraction from Step 5 to the other scale. In this case we want to find the point that is 1/3rd of the way down from the top, on the score side of the interval. The total distance on the score side is 5 points, so the position we want is:

(1/3)(5) = 1.67 points down from the top

FREQUENCY DISTRIBUTIONS

Step 7: Compute the final value by starting at the top of the interval and subtracting the distance you computed in Step 6. In this example, the top of the score interval is 14.5 and we want to come down 1.67 points. The final answer is:

14.50 - 1.67 = 12.83

We have determined that a percentile rank of 50% corresponds to a score of 12.83. Notice that this answer is in agreement with the preliminary estimate of X = 13, that we made in Step 4. (If there is a contradiction between your answer and your estimate, you should check your calculations.)

HINTS AND CAUTIONS

1. When making the list of intervals for a grouped frequency distribution table, some people find it easier to begin the list with the lowest interval and work up to the highest. Fewer mistakes will be made.

2. When interpreting a frequency distribution table, be sure to use both columns, X and f, to get a complete list of the entire set of scores. Remember, the X column does not list all the scores, it simply shows the scale of measurement.

SELF TEST
True/False Questions

1. The number of scores in a distribution can be obtained by adding the frequencies in a frequency distribution table; $N = \Sigma f$.

2. It is customary to list the score categories in a frequency distribution from the highest down to the lowest.

3. In a frequency distribution table, the ΣX may be obtained by simply adding the values in the column labeled X.

4. For a set of n = 30 scores, if the lowest score is X = 15 and the highest score is X = 44, then a grouped frequency distribution table should be used.

5. A distribution of scores is being put into a grouped frequency distribution table with an interval width of 10 points. If the smallest score in the distribution is X = 15, then the bottom interval in the table should be 15-24.

6. In a grouped frequency distribution table, one class interval is identified as 15-19. This interval has a width of four points.

7. If it is appropriate to present a distribution of scores in a polygon, then it would also be appropriate to present the scores in a bar graph.

8. A professor records the academic major for each student in a class. If these data are presented in a frequency distribution graph, the graph should *not* be a polygon.

9. The scores for a very easy exam would probably form a positively skewed distribution.

10. In a grouped frequency distribution table, an interval of 15-19 has a cumulative percentage of cf = 50%. For this distribution, the 50th percentile is X = 19.5.

Multiple-Choice Questions

1. For the following data, N = _____.
    a. 10
    b. 11
    c. 28
    d. cannot be determined from the table

X	f
4	2
3	4
2	3
1	2

2. For the scores in the following table, how many individuals had a score of X = 2?
    a. 1
    b. 2
    c. 3
    d. 4

X	f
4	3
3	5
2	4
1	2

3. For the data in the following table, what is the value of ΣX?
    a. 10
    b. 15
    c. 20
    d. cannot be determined from the table

X	f
4	1
3	2
2	4
1	2

4. A set of scores ranges from a high of X = 67 to a low of X = 23. If these scores are organized in a grouped frequency distribution table with an interval width of 5 points, then roughly how many intervals will be needed?
    a. 13-14
    b. around 9
    c. 44
    d. 45

5. Which types of graphs are used for data from an interval scale?
    a. histograms and bar graphs
    b. polygons and bar graphs
    c. histograms and polygons
    d. histograms, bar graphs, and polygons

6. A distribution of scores is being organized in a grouped frequency distribution table with an interval width of 2 points. If the lowest score in the distribution is X = 41, then the bottom interval in the table should be _____.
    a. 40-42
    b. 41-43
    c. 40-41
    d. 41-42

7. A set of scores ranges from a high of X = 96 to a low of X = 11. If these scores are placed in a grouped frequency distribution table, the best value for the interval width is _____.
    a. 2 points
    b. 5 points
    c. 9 points
    d. 10 points

8. For the following frequency distribution of exam scores, what is the lowest score on the exam?

X	f
90-94	3
85-89	4
80-84	5
75-79	2
70-74	1

    a. X = 70
    b. X = 74
    c. X = 90
    d. cannot be determined

9. For the following frequency distribution of exam scores, how many students had scores higher than X = 79?

X	f
90-94	3
85-89	4
80-84	5
75-79	2
70-74	1

    a. 7
    b. 12
    c. 19
    d. cannot be determined

10. If a distribution of scores is shown in a bar graph, you know that the scores were measured on a _____ scale of measurement.
    a. nominal or ordinal
    b. ordinal or interval
    c. interval or ratio
    d. discrete or continuous

11. The normal distribution is an example of
    a. a histogram showing data from a sample
    b. a polygon showing data from a sample
    c. a bar graph showing data from a population
    d. a smooth curve showing data from a population

12. What shape would you expect for the distribution of scores from a very hard exam for a large class of students?
    a. positively skewed
    b. negatively skewed
    c. symmetrical
    d. normal

13. In a grouped frequency distribution, a cumulative percentage of 80% corresponds to the interval 5-9. For this distribution, the 80th percentile is _____.
    a. 4.5
    b. 5.5
    c. 8.5
    d. 9.5

Questions 14 and 15 refer to the following distribution

X	f	cf	c%
25-29	2	20	100%
20-24	5	18	90%
15-19	7	13	65%
10-14	6	8	40%
5-9	2	2	10%

14. What is the percentile rank corresponding to a score of X = 24.5?
    a. 18%
    b. 90%
    c. 65%
    d. cannot be determined from the information given

15. What is the 40th percentile?
    a. X = 9
    b. X = 9.5
    c. X = 14
    d. X = 14.5

Other Questions

1. Simplifying and organizing data is the goal of descriptive statistics. One descriptive technique is to organize a set of scores in a frequency distribution. Define a frequency distribution and explain how it simplifies and organizes data.

2. Occasionally it is necessary to group scores into class intervals and construct a grouped frequency distribution.
    a. Explain when it is necessary to use a grouped table (as opposed to a regular table).
    b. Outline the guidelines for constructing a grouped frequency table.

3. A frequency distribution graph can be either a histogram, a bar graph, or a polygon. Define each of these graphs and identify the circumstances where each is used.

4. For the following set of scores:
   5, 4, 2, 3, 4, 3, 6, 4, 2, 4, 2, 3, 5, 4, 3
    a. Construct a frequency distribution table including columns for frequency, proportion, and percentage.
    b. Draw a histogram showing the data.

5. For a continuous variable each score actually corresponds to an interval on the scale of measurement.
    a. In general terms define the real limits of an interval.
    b. If a distribution has scores of 10, 9, 8, etc., what are the real limits for $X = 8$?
    c. If a distribution has scores of 5.5, 5.0, 4.5, 4.0, etc., what are the real limits for $X = 4.5$?
    d. In a grouped frequency distribution, each class interval has real limits and apparent limits. What are the real and apparent limits for the interval 10-14?

6. For the following set of scores:
   8, 7, 10, 12, 9, 11, 10, 9, 12, 11
   7, 9, 7, 10, 10, 8, 12, 7, 10, 7
    a. Construct a frequency distribution table including columns for frequency, proportion, and percentage.
    b. Draw a histogram showing the data.

7. Construct a stem-and-leaf display for the following data. What is the shape of the distribution?
   83 46 32 44 75 35 33 47 54 72 60 22
   48 57 49 67 25 51 92 84 43 50 36 40

8. Consider the data in the following table.

X	f	cf	c%
55-59	1	25	100
50-54	2	24	96
45-49	2	22	88
40-44	2	20	80
35-39	4	18	72
30-34	5	14	56
25-29	2	9	36
20-24	4	7	28
15-19	2	3	12
10-14	1	1	4

    a. Find the 50th percentile for this distribution.
    b. Find the 60th percentile.
    c. What is the percentile rank for X = 42?
    d. What is the percentile rank for 44.5?

ANSWERS TO SELF TEST

True/False Answers

1. True
2. True
3. False. To find ΣX you must consider the f values as well as the X values.
4. True
5. False. The bottom number should be a multiple of 10. It should be 10-19.
6. False. The interval width is 5 points.
7. False. A polygon is used for interval or ratio data but a bar graph is for nominal or ordinal scales.
8. True
9. False. The peak is at the right side because most students have high scores.
10. True.

Multiple-Choice Answers

1. b   N is determined by Σf.
2. d   The number of individuals is shown in the f column beside each X.
3. c   Remember to include the frequencies. You must add all 11 scores.
4. b   The scores cover a range of 45 points.
5. c   Interval scale data can be shown in a histogram or a polygon.
6. c   The bottom score is a multiple of the width and each interval contains two scores.
7. d   The range is 86 points.
8. d   The lowest score is in the 70-74 interval but you don't know where.
9. b   Scores in the top three intervals are all at least 80.

10. a   Bar graphs are used for nominal or ordinal scales.
11. d   The normal distribution is a smooth curve used for population data.
12. a   The peak is at the left side because most students have low scores.
13. d   The cumulative percentage is the rank for the upper real limit.
14. b   The cumulative percentage is the rank for the upper real limit.
15. d   The cumulative percentage is the rank for the upper real limit.

Other Answers

1. A frequency distribution shows the number of individuals located in each category on the scale of measurement.

2. a. A grouped frequency distribution table is needed when the range of scores is large, causing a frequency distribution table to have too many entries in the X column.
   b. In a grouped frequency distribution table the guidelines are:
      1. You should have approximately 10 rows in the table.
      2. Interval widths of 2, 5, 10, 20, 50, and 100 should be used. (Select the interval width that satisfies guideline 1.)
      3. The first (lowest) value of each interval should be a multiple of the interval width.
      4. List all intervals without skipping any. The top interval should contain the highest observed X value and the bottom interval should contain the lowest observed X value.

3. In a histogram there is a bar above each score (or interval) showing the frequency. Adjacent bars are touching. A histogram is used with interval or ratio data. A bar graph is similar to a histogram except that there are spaces between the bars and the bar graph is used with nominal or ordinal data. In a polygon, the frequency is indicated by a dot above each score (or interval), and the dots are connected with straight lines. A polygon is used with interval or ratio data.

4. a.
X	f	p	%
6	1	.07	7%
5	2	.13	13%
4	5	.33	33%
3	4	.27	27%
2	3	.20	20%

b.
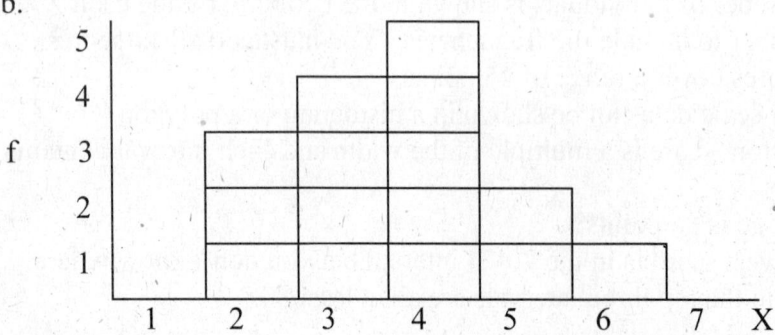

5. a. The real limits for a score are the boundaries located halfway between the score and the next higher (or lower) score.
   b. The real limits for X = 8 would be 7.5 and 8.5.
   c. For X = 4.5, the real limits would be 4.25 and 4.75.
   d. For the class interval 10-14, the real limits are 9.5 and 14.5. The apparent limits are 10 and 14.

6. a.
X	f	p	%
12	3	.15	15%
11	2	.10	10%
10	5	.25	25%
9	3	.15	15%
8	2	.10	10%
7	5	.25	25%

   b.
   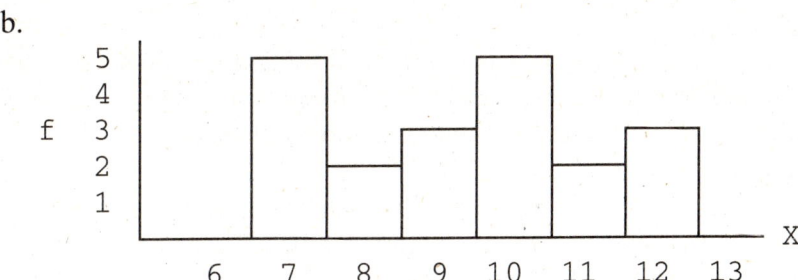

7.
2	25
3	2536
4	6478930
5	4710
6	07
7	52
8	34
9	2

   The distribution is positively skewed.

8. a. X = 33
   b. X = 35.75
   c. c% = 76%
   d. c% = 80%

CHAPTER 3

CENTRAL TENDENCY

CHAPTER SUMMARY

The goals of Chapter 3 are:
1. To introduce central tendency as a descriptive statistical procedure.
2. To introduce the three basic measures of central tendency: the mean, the median, and the mode.
3. To describe how measures of central tendency are related to the shape of the distribution.
4. To describe how central tendency is presented in research reports, including presentation in graphs.

Central Tendency

In general terms, **central tendency** is a statistical measure that determines a single value that accurately describes the center of the distribution and represents the entire distribution of scores. The goal of central tendency is to identify the single value that is the best representative for the entire set of data.

By identifying the "average score," central tendency allows researchers to summarize or condense a large set of data into a single value. Thus, central tendency serves as a descriptive statistic because it allows researchers to describe or present a set of data in a very simplified, concise form. For example, the reading ability for an entire third-grade class can be summarized by the average reading score. In addition, it is possible to compare two (or more) sets of data by simply comparing the average score (central tendency) for one set versus the average score for another set. For example, a report may summarize research results by stating that the patients who received medication had an average cholesterol level 50 points lower than patients without medication.

The Mean, the Median, and the Mode

It is essential that central tendency be determined by an objective and well-defined procedure so that others will understand exactly how the "average" value was obtained and can duplicate the process. However, no single procedure always produces a good, representative value. Therefore, researchers have developed three commonly used techniques for measuring central tendency: the mean, the median, and the mode.

The Mean The mean is the most commonly used measure of central tendency. Computation of the mean requires scores that are numerical values measured on an interval or ratio scale. The mean is obtained by computing the sum, or total, for the entire set of scores, then dividing this sum by the number of scores.

For sample data the mean is: $M = \dfrac{\Sigma X}{n}$

For population data the mean is: $\mu = \dfrac{\Sigma X}{N}$

Conceptually, the mean can also be defined as:
1. The mean is the amount that each individual receives when the total (ΣX) is divided equally among all N individuals.
2. The mean is the balance point of the distribution because the sum of the distances below the mean is exactly equal to the sum of the distances above the mean.

Because the calculation of the mean involves every score in the distribution, changing the value of any score will change the value of the mean. Also, modifying a distribution by discarding scores or by adding new scores will usually change the value of the mean. To determine how the mean will be affected for any specific situation you must consider: 1) how the number of scores is affected, and 2) how the sum of the scores is affected. For example, adding a new score to a distribution will increase the number of scores by 1, and will increase ΣX by the value of the new score.

If a constant value is added to every score in a distribution, then the same constant value is added to the mean. Also, if every score is multiplied by a constant value, then the mean is also multiplied by the same constant value.

Although the mean is the most commonly used measure of central tendency, there are situations where the mean does not provide a good, representative value, and there are situations where you cannot compute a mean at all. When a distribution contains a few extreme scores (or is very skewed), the mean will be pulled toward the extremes (displaced toward the tail). In this case, the mean will not provide a "central" value. With data from a nominal scale it is impossible to compute a mean, and when data are measured on an ordinal scale (ranks), it is usually inappropriate to compute a mean. Thus, the mean does not always work as a measure of central tendency and it is necessary to have alternative procedures available.

The Median If the scores in a distribution are listed in order from smallest to largest, the median is defined as the midpoint of the list. The median divides the scores so that 50% of the scores in the distribution have values that are equal to or less than the median. Computation of the median requires scores that can be placed in rank order (smallest to largest) and are measured on an ordinal, interval, or ratio scale. Usually, the median can be found by a simple counting procedure:
1. With an odd number of scores, list the values in order, and the median is the middle score in the list.
2. With an even number of scores, list the values in order, and the median is half-way between the middle two scores.

If the scores are measurements of a continuous variable, it is possible to find the median by first placing the scores in a frequency distribution histogram with each score represented by a box in the graph. Then, draw a vertical line through the distribution so that exactly half the boxes are on each side of the line. The median is defined by the location of the line.

One advantage of the median is that it is relatively unaffected by extreme scores. Thus, the median tends to stay in the "center" of the distribution even when there are a few extreme scores or when the distribution is very skewed. In these situations, the median serves as a good alternative to the mean.

The Mode The mode is defined as the most frequently occurring category or score in the distribution. In a frequency distribution graph, the mode is the category or score corresponding to the peak or high point of the distribution. The mode can be determined for data measured on any scale of measurement: nominal, ordinal, interval, or ratio.

It is possible for a distribution to have more than one mode. For example, a frequency distribution graph may have two peaks, with a mode at each peak. Such a distribution is called **bimodal**. (Note that a distribution can have only one mean and only one median.) In addition, the term "mode" is often used to describe a peak in a distribution that is not really the highest point. Thus, a distribution may have a *major mode* at the highest peak and a *minor mode* at a secondary peak in a different location.

The primary value of the mode is that it is the only measure of central tendency that can be used for data measured on a nominal scale. In addition, the mode often is used as a supplemental measure of central tendency that is reported along with the mean or the median.

Central Tendency and the Shape of the Distribution

Because the mean, the median, and the mode are all measuring central tendency, the three measures are often systematically related to each other. In a symmetrical distribution, for example, the mean and median will always be equal. If a symmetrical distribution has only one mode, the mode, mean, and median will all have the same value. In a skewed distribution, the mode will be located at the peak on one side and the mean usually will be displaced toward the tail on the other side. The median is usually located between the mean and the mode.

Reporting Central Tendency in Research Reports

In manuscripts and in published research reports, the sample mean is identified with the letter M. There is no standardized notation for reporting the median or the mode. In research situations where several means are obtained for different groups or for different treatment conditions, it is common to present all of the means in a single graph. The different groups or treatment conditions are listed along the horizontal axis and the means are displayed by a bar or a point above each of the groups. The height of the bar (or point) indicates the value of the mean for each group. Similar graphs are also used to show several medians in one display.

LEARNING OBJECTIVES

1. You should be able to define central tendency and you should understand the general purpose of obtaining a measure of central tendency.

2. You should be able to define and compute each of the three basic measures of central tendency for a set of data.

3. You should know how the mean is affected when a set of scores is modified. For example, what happens to the mean when a new score is added to an existing set, or when a score is removed, or when the value of a score is changed. In addition, you should know what happens to the mean when a constant value is added to every score in a distribution, or when every score is multiplied by a constant value.

4. You should know when each of the three measures of central tendency is used and you should understand the advantages and disadvantages of each.

5. You should know how the three measures of central tendency are related to each other for symmetrical distributions and for skewed distributions.

6. You should be able to draw and understand graphs showing the relationship between an independent variable and a dependent variable, where a measure of central tendency (usually the mean) is used to present the "average" score for the dependent variable.

NEW TERMS AND CONCEPTS

The following terms were introduced in this chapter. You should be able to define or describe each term and, where appropriate, describe how each term is related to other terms in the list.

Central tendency	A statistical measures that identifies a single score (usually a central value) to serve as a representative for the entire group.
Mean	The value obtained when the sum of the scores is divided by the number of scores.
Median	The score that divides a distribution exactly in half.
Mode (major and minor)	The score with the greatest frequency overall (major), or the greatest frequency within the set of neighboring scores (minor).

Weighted mean   The average of two means, calculated so that each mean is weighted by the number of scores it represents.

NEW FORMULAS

$$\mu = \frac{\Sigma X}{N}$$

$$M = \frac{\Sigma X}{n}$$

STEP BY STEP

  The Weighted Mean: Occasionally a researcher will find it necessary to combine two (or more) sets of data, or to add new scores to an existing set of data. Rather than starting from scratch to compute the mean for the new set of data, it is possible to compute the weighted mean. To find the new mean you need two pieces of information:
  1) How many scores are in the new data set?
  2) What is the sum of all the scores?
Remember, the mean is the sum of the scores divided by the number. We will use the following problem to demonstrate the calculation of the weighted mean.
  A researcher wants to combine the following three samples into a single group. Notice that sample 3 is actually a single score, $X = 4$. What is the mean for the combined group?

  Sample 1   Sample 2   Sample 3
  $n = 8$ $n = 5$ $n = 1$
  $M = 12$ $M = 9$ $M = 4$

  Step 1: To find n and ΣX for the combined group, the first step is to find n and ΣX for each of the individual samples. For example, sample 1 consists of $n = 8$ scores with a mean of $M = 12$. You can find ΣX for the scores by using the formula for the mean, and substituting the two values that you know,

$$M = \frac{\Sigma X}{n}$$

In this case,

$$12 = \frac{\Sigma X}{8}$$

Multiplying both sides of the equation by 8 gives,

$$8(12) = \Sigma X$$
$$96 = \Sigma X$$

Often this process is easier to understand if you put dollar-signs on the numbers and remember that the mean is the amount that each individual receives if the total is divided equally. For this example, we have a group of n = 8 people and, if the total amount is divided equally, each person will get $12 (M). If the group puts all their money together (ΣX), how much will they have? Again the answer is ΣX = $96.

Step 2: Repeat the process in Step 1 for each individual data set. If the problem involves adding a single score to an existing data set, then you can treat the single score as a sample with n = 1 and X = M = ΣX.

For this example, Sample 2 has a n = 5 and ΣX = 45, and Sample 3 has n = 1 and ΣX = 4.

Step 3: Once you have determined n and ΣX for each individual data set, then you simply add the individual n values to find the number of scores in the combined set. In the same way, you simply add the ΣX values to find the overall sum of the scores in the combined data set. For this example,

combined n = 8 + 5 + 1 = 14
combined ΣX = 96 + 45 + 4 = 145

Step 4: Finally you compute the mean for the combined group using the regular formula for M.

$$M = \Sigma X/n = 145/14 = 10.36$$

HINTS AND CAUTIONS

1. One of the most common errors in computing central tendency occurs when students are attempting to find the mean for data in a frequency distribution table. You must remember that the frequency distribution table condenses a large set of scores into a concise, organized distribution; the table does not list each of the individual scores. One way of avoiding confusion is to transform the frequency distribution table back into the original list of scores. For example, the following frequency distribution table presents a distribution for which three individuals had scores of X = 5, one individual had X = 4, four individuals had X = 3, no one had X = 2, and two individuals had X = 1. When each of these scores is listed individually, it is much easier to see that N = 10 and ΣX = 33 for these data.

Frequency Distribution		Original X Values
X	f	5
5	3	5
4	1	5
3	4	4
2	0	3
1	2	3
		3
		3
		1
		1

2. Many students incorrectly assume that the median corresponds to the midpoint of the range of scores. For example, it is tempting to say that the median for a 100-point test would be X = 50. Be careful! The correct interpretation is that the median divides the set of scores (or individuals) into two equal groups. On a 100-point test, for example, the median could be X = 95, if the test was very easy and 50% of the class scored above 95. You must know where the individual scores are located before you can find the median.

Often it is easy to locate the median if you sketch a histogram of the frequency distribution. If each score is represented by a "block" in the graph, you can find the median by positioning a vertical line so that it divides the blocks into two equal piles.

SELF TEST

True/False Questions

1. A student takes a 10-point quiz each week in statistics class. If the student's quiz scores for the first three weeks are 2, 6, and 10, then the mean score is M = 9.

2. Adding a new score to an existing sample will always change the value of the sample mean.

3. A sample of n = 6 scores has a mean of M = 9. If one individual with a score of X = 4 is removed from the sample, the new mean will be M = 10.

4. A sample of n = 7 scores has a mean of M = 5. After one new score is added to the sample the new mean is calculated to be M = 6. The new score was X = 13.

5. A sample of n = 8 scores has a mean of M = 20. After one score is removed from the sample, the mean is calculated to be M = 23. The removed score must have a value greater than 20.

6. A sample has n = 5 scores: 2, 4, 5, 8, and 11. The median for the sample is 6.5.

7. It is impossible for the value of the mode to be greater than the value of the mean.

8. Occasionally, the median is a more central and representative value than the mean because it is relatively unaffected by extreme scores.

9. The mode is always located in the exact center of a perfectly symmetrical distribution.

10. The mean tends to be displaced toward the tail of a skewed distribution.

Multiple-Choice Questions

1. What is the mean for the population of scores shown in the frequency distribution table?

    a. 15/5 = 3
    b. 15/12 = 1.25
    c. 32/5 = 6.60
    d. 32/12 = 2.67

X	f
5	1
4	2
3	3
2	4
1	2

2. What is the median for the population of scores shown in the frequency distribution table?

    a. 2.5
    b. 3
    c. 3.5
    d. 4

X	f
5	1
4	2
3	3
2	4
1	2

3. What is the mean for the following set of scores?   Scores: 1, 6, 10, 11
    a. 4
    b. 7
    c. 8
    d. 14

4. Which of the following is a property of the mean?
    a. Changing the value of a score will change the value of the mean.
    b. Adding a new score to a distribution will change the value of the mean.
    c. Removing a score from a distribution will change the value of the mean
    d. All 3 of the other choices are correct.

5. A teacher gave a reading test to a class of 5th-grade students and computed the mean, median, and mode for the test scores. Which of the following statements *cannot* be an accurate description of the scores?
   a. No one had a score exactly equal to the mean.
   b. No one had a score exactly equal to the median.
   c. No one had a score exactly equal to the mode.
   d. All of the other options are false statements.

6. A professor records the academic major for each student in a class of n = 40. The best measure of central tendency for these data would be _____.
   a. the mean
   b. the median
   c. the mode
   d. Central tendency cannot be determined for these data.

7. One sample with n = 4 scores has a mean of M = 12, and a second sample with n = 6 scores has a mean of M = 8. If the two samples are combined, what is the mean for the combined set of scores?
   a. 4.8
   b. 9.6
   c. 10.0
   d. 19.2

8. A sample of n = 8 scores has a mean of M = 10. After one score is removed from the sample, the mean for the remaining score is found to be M = 11. What was the score that was removed?
   a. X = 3
   b. X = 7
   c. X = 8
   d. impossible to determine from the information provided

9. A sample of n = 6 scores has a mean of M = 5. One person with a score of X = 12 is added to the distribution. What is the mean for the new set of scores?
   a. M = 5
   b. M = 6
   c. M = 7
   d. M = 8

10. In a sample of n = 6 scores, the smallest score is X = 3, the largest score is X = 10, and the mean is M = 6. If the largest score is changed from X = 10 to X = 22, then what is the value of the new mean?
    a. The mean is still M = 6
    b. The mean is M = 7
    c. The mean is M = 8
    d. impossible to determine from the information given

11. A sample of n = 20 scores has a mean of M = 55. After one score is removed from the sample, the mean for the remaining scores is found to be M = 51. From this information you can conclude that the removed score was _____.
    a. greater than 55
    b. less than 55
    c. equal to 55
    d. It is impossible to estimate the magnitude of the score from the information provided.

12. What is the median for the following set of scores?
    Scores: 1, 3, 9, 10, 22
    a. 6
    b. 9
    c. 9.5
    d. 11

13. For a perfectly symmetrical distribution with a median of 30, what is the value of the mean?
    a. 30
    b. greater than 30
    c. less than 30
    d. cannot be determined from the information given

14. For a positively skewed distribution with $\mu = 30$, what is the most likely location for the median?
    a. equal to 30
    b. greater than 30
    c. less than 30
    d. cannot be determined from the information given

15. Adding one new score located in the extreme right-hand tail of a distribution will
    a. increase the value of the mean
    b. increase the value of the median
    c. increase the value of the mode
    d. increase the value for all three measures of central tendency (mean, median, and mode)

Other Questions

1. Explain the general purpose for obtaining a measure of central tendency.

2. A sample of n = 4 scores has a mean of M = 8.
    a. If a new score, X = 3, is added to the sample, what value will be obtained for the new sample mean?

b. If one of the scores, X = 5, is removed from the original sample, what value would be obtained for the new sample mean?
c. If one of the scores in the original sample is changed from X = 10 to X = 18, what value would be obtained for the new sample mean?

3. One sample has n = 6 and M = 10. A second sample has n = 4 scores with M = 15. If the two samples are combined, what is the value of the mean for the combined sample?

4. Compute the mean, median, and mode for the following set of scores.
   Scores: 5, 7, 5, 4, 3, 12, 9, 6, 6, 5, 7, 5, 6, 4

5. Compute the mean, median, and mode for the set of scores shown in the following frequency distribution table.

X	f
7	1
6	2
5	3
4	4
3	1
2	0
1	1

6. A researcher obtains the following results from an experiment comparing three treatment conditions:
   Treatment #1: M = 12.7
   Treatment #2: M = 20.5
   Treatment #3: M = 8.4
   a. Assuming that the independent variable (the differences between treatments) is measured on a nominal scale, sketch a graph showing the experimental results.
   b. Assuming that the independent variable is measured on an interval scale, sketch a graph showing the results.

ANSWERS TO SELF TEST

True/False Answers

1. False. The mean is 18/3 = 6.
2. False Not if the score equals the mean.
3. True
4. True
5. False. The score must be less than 20.
6. False The median is 5.
7. False. The mode is not necessarily greater than or less than the mean.

8. True
9. False Not if the distribution is bimodal.
10. True

Multiple-Choice Answers

1. d You must use the frequencies as well as the X values.
2. a The 12 scores are divided into two groups of 6..
3. b The mean is 28/4 = 7.
4. a Changing a score always changes the mean.
5. c The mode is the most commonly occurring score. There always will be scores located at the mode.
6. c The mode must be used for scores from a nominal scale.
7. b The two samples have a total of 10 scores that sum to 96.
8. a The original sample has a sum of 80 for 8 scores and the new sample has a sum of 77 for 7 scores.
9. b The original scores sum to 30. The new sum is 42 for n = 7 scores.
10. c Adding 12 points to one score adds 12 points to the total which increases from 36 to 48. The new mean is 48/6 = 8.
11. a Removing a large score causes the mean to drop.
12. b With an odd number, the median is the middle score.
13. a In a perfectly symmetrical distribution the mean equals the median.
14. c The mean is displaced toward the tail.
15. a The other two measures may be affected but the mean will definitely increase.

Other Answers

1. The purpose of central tendency is to find a single value that best represents an entire distribution of scores.

2. a. With n = 4 and M = 8, the original sample has a total of $\Sigma X = 32$. Adding X = 3 produces n = 5 and $\Sigma X = 35$. The new mean is 35/5 = 7.
   b. With n = 4 and M = 8, the original sample has a total of $\Sigma X = 32$. Taking away X = 5 leaves a sample with n = 3 scores and $\Sigma X = 27$. The new mean is 27/3 = 9.
   c. Changing X = 10 to X = 18 adds 8 points to the total but does not change the number of scores. The new sample has n = 4 with $\Sigma X = 40$. The new mean is 40/4 = 10.

3. The first sample has n = 6, M = 10, and $\Sigma X = 60$. The second sample has n = 4, M = 15, and $\Sigma X = 600$. When the samples are combined, the total number of score is n = 10 and the sum of the scores is $\Sigma X = 120$. The mean for the combined sample is 120/10 = 12.

4. The mean is 84/14 = 6.00. The median is X = 5.5, and the mode is X = 5.

5. The mean is 54/12 = 4.5. The median is X = 4.5, and the Mode is X = 4.

6. a.

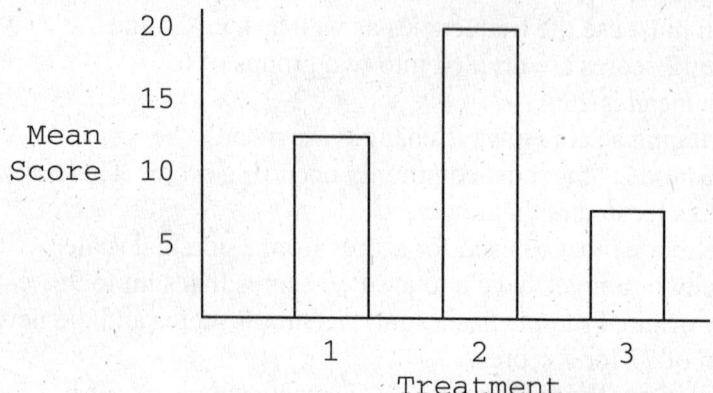

b. Use a histogram or a line graph (shown as follows).

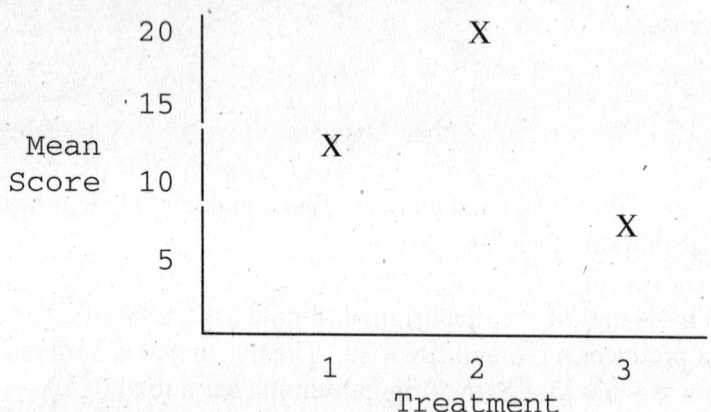

CHAPTER 4

VARIABILITY

CHAPTER SUMMARY

The goals of Chapter 4 are:
1. To introduce the concept of variability and to describe its importance as a descriptive measure and as a component of inferential statistics.
2. To introduce the basic techniques for measuring and describing variability: the range, the interquartile range, and the standard deviation/variance.
3. To demonstrate the properties of the standard deviation.
4. To demonstrate how the mean and standard deviation, working together, can provide a detailed description for a set of scores.

Variability

The goal for variability is to obtain a measure of how spread out the scores are in a distribution. A measure of variability usually accompanies a measure of central tendency as basic descriptive statistics for a set of scores. Central tendency describes the central point of the distribution, and variability describes how the scores are scattered around that central point. Together, central tendency and variability are the two primary values that are used to describe a distribution of scores.

Variability serves both as a descriptive measure and as an important component of most inferential statistics. As a descriptive statistic, variability measures the degree to which the scores are spread out or clustered together in a distribution. In the context of inferential statistics, variability provides a measure of how accurately any individual score or sample represents the entire population. When the population variability is small, all of the scores are clustered close together and any individual score or sample will necessarily provide a good representation of the entire set. On the other hand, when variability is large and scores are widely spread, it is easy for one or two extreme scores to give a distorted picture of the general population.

Measuring Variability: The Range, the Interquartile Range, and the Standard Deviation/Variance

Variability can be measured with the range, the interquartile range, or the standard deviation/variance. In each case, variability is determined by measuring *distance*. The **range** is the total distance covered by the distribution, from the highest score to the lowest score (using the upper and lower real limits of the range). The **interquartile range** is the distance covered by the middle 50% of the distribution (the difference between Q1 and Q3). **Standard deviation** measures the standard distance between a score and the mean. The calculation of standard deviation can be summarized as a four-step process:
1. Compute the deviation (distance from the mean) for each score.
2. Square each deviation.

3. Compute the mean of the squared deviations. For a population, this involves summing the squared deviations (sum of squares, SS) and then dividing by N. The resulting value is called the **variance** or *mean square* and measures the average squared distance from the mean.

   For samples, variance is computed by dividing the sum of the squared deviations (SS) by n - 1, rather than N. The value, n - 1, is know as degrees of freedom (df) and is used so that the sample variance will provide an unbiased estimate of the population variance.

4. Finally, take the square root of the variance to obtain the standard deviation.

Properties of the Standard Deviation

Because the standard deviation is probably the most commonly used measure of variability, it is important to understand how this descriptive measure is affected when a set of scores is transformed by either adding a constant or multiplying each score by a constant. If a constant is added to every score in a distribution, the standard deviation will *not* be changed. If you visualize the scores in a frequency distribution histogram, then adding a constant will move each score so that the entire distribution is shifted to a new location. The center of the distribution (the mean) changes, but the standard deviation remains the same. If each score is multiplied by a constant, the standard deviation will be multiplied by the same constant. Multiplying by a constant will multiply the distance between scores, and because the standard deviation is a measure of distance, it will also be multiplied.

The Mean and Standard Deviation as Descriptive Statistics

If you are given numerical values for the mean and the standard deviation, you should be able to construct a visual image (or a sketch) of the distribution of scores. For example, a set of scores with $M = 80$ and $s = 10$ will form a distribution that is centered at $M = 80$, with most of the scores located between 70 and 90 (within 10 points of the mean). As a general rule, about 70% of the scores will be within one standard deviation of the mean, and about 95% of the scores will be within a distance of two standard deviations of the mean.

LEARNING OBJECTIVES

1. Understand the measures of variability and be able to tell the difference between sets of scores with low versus high variability.

2. Know how to calculate SS (the sum of squared deviations) using either the computational or definitional formula.

3. Be able to calculate the population and sample variance and standard deviation, and understand the correction used in the formula for sample variance.

4. Be familiar with the characteristics of measures of variability, especially those for standard deviation.

NEW TERMS AND CONCEPTS

The following terms were introduced in this chapter. You should be able to define or describe each term and, where appropriate, describe how each term is related to other terms in the list.

Variability	A measure of the degree to which the scores in a distribution are clustered together or spread apart.
Range	The distance from the upper real limit of the highest score to the lower real limit of the lowest score; the total distance from the absolute highest point to the lowest point in the distribution.
First quartile	The score that separates the lowest 25% of a distribution from the highest 75%.
Third quartile	The score that separates the lowest 75% of a distribution from the highest 25%.
Interquartile range	The distance from the first quartile to the third quartile.
Deviation score	The distance (and direction) from the mean to a specific score. Deviation = $X - \mu$.
Population variance	The average squared distance from the mean; the mean of the squared deviations.
Population standard deviation	The square root of the population variance; a measure of the standard distance from the mean.
Sample variance	The sum of the squared deviations divided by df = $n - 1$. An unbiased estimate of the population variance.
Sample standard deviation	The square root of the sample variance.
Degrees of freedom	The number of scores in a sample that are independent and free to vary with no restriction. df = $n - 1$.
Biased statistic	A statistic that, on average, consistently tends to overestimate (or underestimate) the corresponding population parameter.
Unbiased statistic	A statistic that, on average, provides an accurate estimate of the corresponding population parameter. The sample mean and sample variance are unbiased statistics.

NEW FORMULAS

For a population:

$$SS = \Sigma(X - \mu)^2 \quad \text{or} \quad SS = \Sigma X^2 - \frac{(\Sigma X)^2}{N}$$

$$\sigma^2 = \frac{SS}{N} \qquad \sigma = \sqrt{\frac{SS}{N}}$$

For a sample:

$$SS = \Sigma(X - M)^2 \quad \text{or} \quad SS = \Sigma X^2 - \frac{(\Sigma X)^2}{n}$$

$$s^2 = \frac{SS}{n-1} \qquad s = \sqrt{\frac{SS}{n-1}} \qquad df = n - 1$$

STEP BY STEP

<u>SS, Variance, and Standard Deviation</u>: The following set of data will be used to demonstrate the calculation of SS, variance, and standard deviation.
Scores: 5, 3, 2, 4, 1

Step 1: Before you begin any calculation, simply look at the set of scores and make a preliminary estimate of the mean and standard deviation. For this set of data, it should be obvious that the mean is around 3, and most of the scores are within one or two points of the mean. Therefore, the standard deviation (standard distance from the mean) should be about 1 or 2.

Step 2: Determine which formula you will use to compute SS. If you have a relatively small set of data and the mean is a whole number, then use the definitional formula. Otherwise, the computational formula is a better choice. For this example there are only 5 scores and the mean is equal to 3. The definitional formula would be fine for these scores.

Step 3: Calculate SS. Note that it does not matter whether the set of scores is a sample or a population when you are computing SS. For this example, we will use formulas with population notation, but using sample notation would not change the result.

Definitional Formula: List each score in a column. In a second column put the deviation score for each X value. (Check that the deviations add to zero). In a third column list the squared deviation scores. Then simply add the values in the third column.

X	(X - μ)	(X - μ)²
5	2	4
3	0	0
2	−1	1
4	1	1
1	−2	4
	0	10 = SS

Computational Formula: List each score in a column. In a second column list the squared value for each X. Then find the sum for each column. These are the two sums that are needed for the computational formula.

X	X²
5	25
3	9
2	4
4	16
1	1
15	55

$\Sigma X = 15$

$\Sigma X^2 = 55$

Then use the two sums in the computational formula to calculate SS.

$$SS = \Sigma X^2 - \frac{(\Sigma X)^2}{N} = 55 - (15)^2/5 = 55 - 45 = 10$$

Step 4: Now you must determine whether the set of scores is a sample or a population. With a population, you use N in the formulas for variance and standard deviation. With a sample, use n − 1.

For a Population

$\sigma^2 = SS/N$
$= 10/5 = 2$

$\sigma = \sqrt{2} = 1.41$

For a Sample

$s^2 = SS/(n-1)$
$= 10/4 = 2.5$

$s = \sqrt{2.5} = 1.58$

HINTS AND CAUTIONS

1. Mistakes are commonly made in the computational formula for SS. Often, ΣX^2 is confused with $(\Sigma X)^2$. The ΣX^2 indicates that the X values are first squared, then summed. On the other hand, $(\Sigma X)^2$ requires that the X values are first added, then the sum is squared.

VARIABILITY

2. Remember, it is impossible to get a negative value for SS because, by definition, SS is the sum of the squared deviation scores. The squaring operation eliminates all of the negative signs.

3. The computational formula is usually easier to use than the definitional formula because the mean usually is not a whole number.

4. When computing a variance or a standard deviation, be sure to check whether you are computing the measure for a population or a sample. Remember, the sample variance uses n – 1 in the denominator so that it will provide an unbiased estimate of the population variance.

5. Note that you do not use n – 1 in the formula for sample SS. The value n – 1 is used to compute sample variance *after* you have calculated SS.

SELF TEST

True/False Questions

1. The range, the interquartile range, and the standard deviation, are all measures of distance.

2. If the scores in a sample range from a high of X = 16 to a low of X = 9, then the range is 8 points.

3. The first quartile (Q1) is the score that separates the top 25% of a distribution from the bottom 75%.

4. A negative deviation always indicates a score located below the mean.

5. The value of SS will always be greater than or equal to zero.

6. If the population variance is 4, then the standard deviation will be σ = 16.

7. A population has a mean of μ = 50 and σ = 10. If 5 points are added to every score in the population, the new mean and standard deviation will be μ = 55 and σ = 15.

8. A population has a mean of μ = 20 and σ = 4. If every score in the population is multiplied by 2, the new mean and standard deviation will be μ = 40 and σ = 8.

9. A sample of n = 7 scores has SS = 42. The variance for this sample is s^2 = 6.

10. In a population with a mean of μ = 40 and a standard deviation of σ = 8, a score of X = 46 would be an extreme value, far out in the tail of the distribution.

Multiple-Choice Questions

1. For the following sample of n = 8 scores, what is the value of the range?
   Sample: 2, 3, 4, 4, 5, 7, 8, 10
   a. 2 points
   b. 4 points
   c. 8 points
   d. 9 points

2. For the following sample of n = 8 scores, what is the value of the semi-interquartile range?
   Sample: 2, 3, 4, 4, 5, 7, 8, 10
   a. 2 points
   b. 4 points
   c. 8 points
   d. 9 points

3. What is the value of SS for the following set of scores? Scores: 5, 6, 1.
   a. 144
   b. 62
   c. 14
   d. None of the other choices is correct.

4. What is the value of SS for the following set of scores? Scores: 8, 3, 1.
   a. 26
   b. 29
   c. 74
   d. 144

5. A population of N = 5 scores has $\Sigma X = 20$ and $\Sigma X^2 = 100$. For this population, what is the value of SS?
   a. 20
   b. 80
   c. 100
   d. 380

6. A sample of n = 5 scores has SS = 40. If the same five scores were a population, then the value obtained for the population SS would be _____.
   a. 40
   b. 5(40/4) = 50
   c. 4(40/5) = 32
   d. 5(40) = 200

7. A population of N = 10 scores has μ = 50 and SS = 200. For this population, what is the value of Σ(X − μ)?
   a. 0
   b. √200
   c. 450
   d. cannot be determined from the information given

8. A population of N = 10 scores has μ = 50 and SS = 200. For this population, what is the value of Σ(X − μ)²?
   a. 0
   b. 200
   c. (450)²
   d. cannot be determined from the information given

9. A sample of n = 10 scores is selected from a population. The variability of the scores in the sample will tend to be _____ the variability of the scores in the population.
   a. less than
   b. greater than
   c. equal to

10. A set of 10 scores has SS = 90. If the scores are a sample, the sample variance is ____ and if the scores are a population, the population variance is ____.
    a. $s^2 = 9$, $\sigma^2 = 9$
    b. $s^2 = 9$, $\sigma^2 = 10$
    c. $s^2 = 10$, $\sigma^2 = 9$
    d. $s^2 = 10$, $\sigma^2 = 10$

11. Without doing any serious calculations, which of the following samples has the largest variance?
    a. 1, 3, 4, 5, 6
    b. 1, 5, 8, 12, 22
    c. 30, 32, 34, 35, 36

12. A population has μ = 50 and σ = 5. If 10 points are added to every score in the population, then what are the new values for the mean and standard deviation?
    a. μ = 50 and σ = 5
    b. μ = 50 and σ = 15
    c. μ = 60 and σ = 5
    d. μ = 60 and σ = 15

13. A population of scores has μ = 50 and σ = 5. If every score in the population is multiplied by 3, then what are the new values for the mean and standard deviation?
    a. μ = 50 and σ = 5
    b. μ = 50 and σ = 15
    c. μ = 150 and σ = 5
    d. μ = 150 and σ = 15

14. Which of the following is true for most distributions?
    a. Around 30% of the scores will be located within one standard deviation of the mean.
    b. Around 50% of the scores will be located within one standard deviation of the mean.
    c. Around 70% of the scores will be located within one standard deviation of the mean.
    d. Around 90% of the scores will be located within one standard deviation of the mean.

15. If you have a score of X = 75 on an exam, which set of parameters would give you the highest position within the class?
    a. $\mu = 70$ and $\sigma = 5$
    b. $\mu = 70$ and $\sigma = 10$
    c. $\mu = 60$ and $\sigma = 5$
    d. $\mu = 60$ and $\sigma = 10$

Other Questions

1. Calculate the variance and standard deviation for the following sample of scores:
    7, 2, 4, 6, 4, 7, 3, 7

2. Compute the variance and standard deviation for the following population of scores:
    1, 9, 8, 5, 7

3. Calculate SS, variance, and standard deviation for the following sample of scores:
    15, 16, 9, 1, 9.

4. Describe what happens to the deviation scores and the standard deviation when a constant is added to every score in the distribution.

ANSWERS TO SELF TEST
True/False Answers

1. True
2. True
3. False. Q1 separates the bottom 25% from the top 75%.
4. True
5. True
6. False. Standard deviation is the square root of the variance.
7. False. The standard deviation is still 10.
8. True
9. False. Divide SS by n – 1 for sample variance.
10. False. The score is less than 1 standard deviation above the mean.

Multiple-Choice Answers

1. d For whole number scores, the range is equal to the high score minus the low score, plus 1.
2. a Q1 = 3.5 and Q3 = 7.5.
3. c The deviations are 1, 2, and –3. The squared values sum to 14.
4. a The deviations are 4, –1, and –3. The squared values sum to 26.
5. a SS equals ΣX^2 minus $(\Sigma X)^2$ divided by n.
6. a SS is computed the same way for a sample or a population.
7. a $\Sigma(X - \mu)$ is the sum of the deviations and is always equal to zero.
8. b $\Sigma(X - \mu)^2$ is the sum of the squared deviations (SS).
9. a Sample variability is biased and underestimates the population value.
10. c For a sample, divide by n – 1. For a population, divide by N.
11. b Variance measures how spread out the scores are.
12. c Adding a constant will cause the mean to change but has no effect on the standard deviation.
13. d Multiplying by a constant multiplies both the mean and the standard deviation.
14. c Roughly 70% are within one standard deviation and 95% are within two.
15. c Your score is three times the standard distance above the mean.

Other Answers

1. SS = 28, n = 8, df = 7, s^2 = 4, and s = 2

2. SS = 40, n = 5, σ^2 = 8, σ = 2.83

3. SS = 144, s^2 = 36, and s = 6

4. If a constant is added to each score, the mean also is increased by that constant. The deviation scores are not changed. If the deviation scores have not changed, then the squared deviations and SS will be unchanged. Thus, when a constant is added to every score in a distribution, the standard deviation is not changed.

CHAPTER 5
z-SCORES

CHAPTER SUMMARY

The goals of Chapter 5 are:
1. To introduce z-scores as a method for describing exact locations in a distribution.
2. To demonstrate how raw scores (X values) are transformed into z-scores, and how to reverse the transformation.
3. To demonstrate the concept of standardizing a distribution by transforming all of the scores into z-scores.
4. To demonstrate how z-scores can be used to create a standardized distribution with any predetermined values for the mean and standard deviation.

z-Scores and Location

By itself, a raw score or X value provides very little information about how that particular score compares with other values in the distribution. A score of X = 53, for example, may be a relatively low score, or an average score, or an extremely high score depending on the mean and standard deviation for the distribution from which the score was obtained. If the raw score is transformed into a z-score, however, the value of the z-score tells exactly where the score is located relative to all the other scores in the distribution. The process of changing an X value into a z-score involves creating a signed number, called a **z-score**, such that
  a. The sign of the z-score (+ or −) identifies whether the X value is located above the mean (positive) or below the mean (negative).
  b. The numerical value of the z-score corresponds to the number of standard deviations between X and the mean of the distribution.

Thus, a score that is located two standard deviations above the mean will have a z-score of +2.00. And, a z-score of +2.00 always indicates a location above the mean by two standard deviations.

Transforming back and forth between X and z

The basic z-score definition is usually sufficient to complete most z-score transformations. However, the definition can be written in mathematical notation to create a formula for computing the z-score for any value of X.

$$z = \frac{X - \mu}{\sigma}$$

Also, the terms in the formula can be regrouped to create an equation for computing the value of X corresponding to any specific z-score.

$$X = \mu + z\sigma$$

In addition to knowing the basic definition of a z-score and the formula for a z-score, it is useful to be able to visualize z-scores as locations in a distribution (see Figure 5.2 in the text). Remember, z = 0 is in the center (at the mean), and the extreme tails correspond to z-scores of approximately –2.00 on the left and +2.00 on the right. Although more extreme z-score values are possible, most of the distribution is contained between z = –2.00 and z = +2.00.

The fact that z-scores identify exact locations within a distribution, means that z-scores can be used as descriptive statistics and as inferential statistics. As descriptive statistics, z-scores describe exactly where each individual is located. As inferential statistics, z-scores determine whether a specific sample is representative of its population, or is extreme and unrepresentative. For example, a sample with a z-score near zero is a central, typical sample located near the population mean. On the other hand, a sample with a z-score value beyond 2.00 (or –2.00) would be considered an extreme or unusual sample, much different from the population mean.

z-Scores as a Standardized Distribution
When an entire distribution of X values is transformed into z-scores, the resulting distribution of z-scores will always have a mean of zero and a standard deviation of one. The transformation does not change the shape of the original distribution and it does not change the location of any individual score relative to others in the distribution.
The advantage of standardizing distributions is that two (or more) different distributions can be made the same. For example, one distribution has $\mu = 100$ and $\sigma = 10$, and another distribution has $\mu = 40$ and $\sigma = 6$. When these distribution are transformed to z-scores, both will have $\mu = 0$ and $\sigma = 1$. Because z-score distributions all have the same mean and standard deviation, individual scores from different distributions can be directly compared. A z-score of +1.00 specifies the same location in all z-score distributions.

z-Scores and Samples
It is also possible to calculate z-scores for samples. The definition of a z-score is the same for either a sample or a population, and the formulas are also the same except that the sample mean and standard deviation are used in place of the population mean and standard deviation. Thus, for a score from a sample,

$$z = \frac{X - M}{s}$$

Using z-scores to standardize a sample also has the same effect as standardizing a population. Specifically, the mean of the z-scores will be zero and the standard deviation of the z-scores will be equal to 1.00 provided the standard deviation is computed using the sample formula (dividing n – 1 instead of n).

Other Standardized Distributions Based on z-Scores
Although transforming X values into z-scores creates a standardized distribution, many people find z-scores burdensome because they consist of many decimal values and

negative numbers. Therefore, it is often more convenient to standardize a distribution into numerical values that are simpler than z-scores. To create a simpler standardized distribution, you first select the mean and standard deviation that you would like for the new distribution. Then, z-scores are used to identify each individual's position in the original distribution and to compute the individual's position in the new distribution. Suppose, for example, that you want to standardize a distribution so that the new mean is $\mu = 50$ and the new standard deviation is $\sigma = 10$. An individual with $z = -1.00$ in the original distribution would be assigned a score of $X = 40$ (below μ by one standard deviation) in the standardized distribution. Repeating this process for each individual score allows you to transform an entire distribution into a new, standardized distribution.

LEARNING OBJECTIVES

1. You should be able to describe and understand the purpose for z-scores.

2. You should be able to transform X values into z-scores or transform z-scores into X values.

3. You should be able to describe the effects of standardizing a distribution by transforming the entire set of raw scores into z-scores.

4. Using z-scores, you should be able to transform any set of scores into a distribution with a predetermined mean and standard deviation.

NEW TERMS AND CONCEPTS

The following terms were introduced in this chapter. You should be able to define or describe each term and, where appropriate, describe how each term is related to other terms in the list.

Raw score	An original, untransformed observation or measurement.
z-score	A standardized score with a sign that indicates direction from the mean (+ above μ and – below μ), and a numerical value equal to the distance from the mean measured in standard deviations.
z-score transformation	A transformation that changes raw scores (X values) into z-scores.
Standard score	A score that has been transformed into a standard form.
Standardized distribution	An entire distribution that has been transformed to create predetermined values for μ and σ.

NEW FORMULAS

For a population:

$$z = \frac{X - \mu}{\sigma}$$

$$X = \mu + z\sigma$$

For a sample:

$$z = \frac{X - M}{s}$$

$$X = M + zs$$

STEP BY STEP

Changing X to z: The process of changing an X value to a z-score involves finding the precise location of X within its distribution. We will begin with a distribution with $\mu = 60$ and $\sigma = 12$. The goal is to find the z-score for $X = 75$.

Step 1: First determine whether X is above or below the mean. This will determine the sign of the z-score. For our example, X is above μ so the z-score will be positive.

Step 2: Next, find the distance between X and μ. For our example,
$X - \mu = 75 - 60 = 15$ points

Note: Steps 1 and 2 simply determine a deviation score (sign and magnitude). If you are using the z-score formula, these two steps correspond to the numerator of the equation.

Step 3: Convert the distance from Step 2 into standard deviation units. In the z-score equation, this step corresponds to dividing by σ. For this example,
$15/12 = 1.25$

If you are using the z-score definition (rather than the formula), you simply compare the magnitude of the distance (Step 2) with the magnitude of the standard deviation. For this example, our distance of 15 points is equal to one standard deviation plus 3 more points. The extra 3 points are equal to one-quarter of a standard deviation, so the total distance is one and one-quarter standard deviations.

Step 4: Combine the sign from Step 1 with the number of standard deviations you obtained in Step 3. For this example,
$z = +1.25$

Changing z to X: The process of converting a z-score into an X value corresponds to finding the score that is located at a specified position in a distribution. Again, suppose we have a population with $\mu = 60$ and $\sigma = 12$. What is the X value corresponding to $z = -0.50$?

Step 1: The sign of the z-score tells whether X is above or below the mean. For this example, the X value we want is below μ.

Step 2: The magnitude of the z-score tells how many standard deviations there are between X and μ. For this example, the distance is one-half a standard deviation which is (1/2)(12) = 6 points.

Step 3: Starting with the value of the mean, use the direction (Step 1) and the distance (Step 2) to determine the X value. For this example, we want to find the score that is 6 points below μ = 60. Therefore,
X = 60 − 6 = 54

HINTS AND CAUTIONS

1. Rather than memorizing formulas for z-scores, we suggest that you rely on the definition of a z-score. Remember a z-score identifies a location by specifying the direction from the mean (+ or −) and the distance from the mean in terms of standard deviations.

2. When transforming scores from X to z (or from z to X) it is wise to check your answer by reversing the transformation. For example, given a population with μ = 54 and σ = 4 a score of X = 46 corresponds to a z-score of

$$z = \frac{X - \mu}{\sigma} = \frac{46 - 54}{4} = \frac{-8}{4} = -2.00$$

To check this answer, convert the z-score back into an X value. In this case, z = −2.00 specifies a location below the mean by 2 standard deviations. This distance is
zσ = −2.00(4) = −8 points

With a mean of μ = 54, the score must be

X = 54 − 8 = 46.

SELF TEST

True/False Questions

1. For a population with a mean of μ = 80, any score greater than 80 will have a positive z-score.

2. For a population with a standard deviation of σ = 12, a z-score of z = +0.50 corresponds to a score that is above the mean by 6 points.

3. For a population with a mean of μ = 80 and a standard deviation of σ = 12, a score of X = 77 corresponds to z = –0.50.

4. For a sample with a mean of M = 50 and a standard deviation of s = 10, a z-score of z = +2.00 corresponds to X = 70.

5. For a population with a mean of μ = 40, a score of X = 37 corresponds to z = –0.50. The standard deviation for the population is σ = 3.

6. For a sample with a standard deviation of s = 8, a score of X = 42 corresponds to z = –0.25. The mean for the sample is M = 40.

7. If your exam score is below the mean by 10 points, you would expect a better grade if σ = 5 than if σ = 10.

8. If an entire set of scores (X values) is transformed into z-scores, the set of z-scores will have a mean of zero.

9. One reason for transforming X values into z-scores is that the set of z-scores will form a normal shaped distribution.

10. A population with μ = 45 and σ = 8 is standardized to create a new distribution with μ = 100 and σ = 20. In this transformation, a score of X = 41 from the original distribution will be transformed into a score of X = 110.

Multiple-Choice Questions

1. For a population with μ = 80 and σ = 10, what is the z-score corresponding to X = 95?
    a. +0.25
    b. +0.50
    c. +0.75
    d. +1.50

2. For a sample with M = 50 and s = 12, what is the X value corresponding to z = –0.25?
    a. 47
    b. 53
    c. 46
    d. 54

3. A z-score of z = –2.00 indicates a position
    a. below the mean by 2 points
    b. below the mean by 2 times the standard deviation
    c. above the mean by 2 points
    d. above the mean by 2 times the standard deviation

4. For a population with a standard deviation of $\sigma = 6$, what is the z-score corresponding to a score that is 12 points above the mean?
    a. z = 1
    b. z = 2
    c. z = 6
    d. z = 12

5. For a population with a mean of $\mu = 100$, what is the z-score corresponding to a score that is located 10 points below the mean?
    a. +1
    b. –1
    c. –10
    d. cannot answer without knowing the standard deviation

6. For a population with $\sigma = 10$, a score of X = 60 corresponds to z = –1.50. What is the population mean?
    a. 30
    b. 45
    c. 75
    d. 90

7. For a sample with M = 80, a score of X = 88 corresponds to z = 2.00. What is the sample standard deviation?
    a. 2
    b. 4
    c. 8
    d. 16

8. Under what circumstances would a score that is 15 points above the mean be considered an extreme score, far out in the tail of the distribution?
    a. when the population mean is much larger than 15
    b. when the population standard deviation is much larger than 15
    c. when the population mean is much smaller than 15
    d. when the population standard deviation is much smaller than 15

9. A population with μ = 85 and σ = 12 is transformed into z-scores. After the transformation, what are the values for the mean and standard deviation for the population of z-scores?
    a. μ = 85 and σ = 12
    b. μ = 0 and σ = 12
    c. μ = 85 and σ = 1
    d. μ = 0 and σ = 1

10. A population has μ = 50. What value of σ would make X = 55 central, representative score in the population?
    a. σ = 1
    b. σ = 5
    c. σ = 10
    d. cannot determine with the information given

11. You have a score of X = 65 on an exam. Which set of parameters would give you the best grade on the exam?
    a. μ = 60 and σ = 10
    b. μ = 60 and σ = 5
    c. μ = 70 and σ = 10
    d. μ = 70 and σ = 5

12. For an exam with a mean of M = 74 and a standard deviation of s = 8, Mary has a score of X = 80, Bob's score corresponds to z = +1.50, and Sue's score is located above the mean by 10 points. If the students are placed in order from smallest score to largest score, what is the correct order?
    a. Bob, Mary, Sue
    b. Sue, Bob, Mary
    c. Mary, Bob, Sue
    d. Mary, Sue, Bob

13. Which of the following is an advantage of transforming X values into z-scores?
    a. All negative numbers are eliminated.
    b. The distribution is transformed to a normal shape.
    c. All scores are moved closer to the mean.
    d. None of the other options is an advantage.

14. A distribution with μ = 55 and σ = 6 is being standardized so that the new mean and standard deviation will be μ = 50 and σ = 10. When the distribution is standardized, what value will be obtained for a score of X = 52 from the original distribution?
    a. X = 45
    b. X = 47
    c. X = 52
    d. X = 58

15. A distribution with μ = 35 and σ = 8 is being standardized so that the new mean and standard deviation will be μ = 50 and σ = 10. In the new, standardized distribution your score is X = 60. What was your score in the original distribution?
  a. X = 45
  b. X = 43
  c. X = 1.00
  d. impossible to determine without more information

Other Questions

1. For a population with μ = 90 and σ = 25, find the z-score corresponding to each of the following X values.
  a. X = 95
  b. X = 110
  c. X = 65
  d. X = 80

2. For a sample with M = 60 and s = 6, find the X value corresponding to each of the following z-scores.
  a. z = +1.50
  b. z = –0.50
  c. z = +2.00
  d. z = –1/3

3. On an exam with μ = 70 and σ = 10, you have a score of X = 85.
  a. What is your z-score on this exam?
  b. If the instructor added 5 points to every score, what would happen to your z-score?
  c. If the instructor multiplied every score by 2, what would happen to your z-score?

4. A set of exam scores has μ = 48 and σ = 8. The instructor would like to transform the scores into a standardized distribution with μ = 100 and σ = 20. Find the transformed value for each of the following scores from the original population.
  a. X = 48
  b. X = 50
  c. X = 44
  d. X = 32

ANSWERS TO SELF TEST

True/False Answers

1. True
2. True
3. False. X = 77 corresponds to z = –0.25.
4. True
5. False. The standard deviation is σ = 6.
6. False. The score is below the mean: M = 44.
7. False. You are closer to the mean if σ = 10.
8. True
9. False. Changing X to z does not change the shape of the distribution.
10. False. X = 41 is transformed into X = 90.

Multiple-Choice Answers

1. d X = 95 is above the mean (+) by exactly 1.50 standard deviations.
2. a X = 47 is below the mean by 0.25 standard deviations.
3. b The z-score identifies the direction and distance from the mean in terms of the number of standard deviations.
4. b 12 points is exactly 2 standard deviations.
5. d You cannot compute a z-score without knowing the standard deviation.
6. c 1.5 standard deviations is 15 points below the mean.
7. b If 8 points is two standard deviations, then σ = 4.
8. d 15 points is a small distance when the standard deviation is more than 15.
9. d z-scores always have a mean of zero and a standard deviation of one.
10. c With a standard deviation of 10, X = 55 is only slightly above the mean.
11. b μ = 60 and σ = 5 produce the largest z-score.
12. d z = 0.75 for Mary, z = 1.25 for Sue, z = 1.50 for Bob
13. d None of these characteristics is associated with z-scores.
14. a X = 52 corresponds to z = –0.50. In the new distribution this location corresponds to X = 45.
15. b X = 60 corresponds to z = 1.00. In the old distribution this location corresponds to X = 43.

Other Answers

1. a. z = +0.20
   b. z = +0.80
   c. z = –1.00
   d. z = –0.40

2. a. X = 69
   b. X = 57
   c. X = 72
   d. X = 58

3. a. Your z-score is z = 1.50.
   b. Adding 5 points to every score would increase your score and the mean by 5 points. However, your z-score (your position within the distribution) would not change.
   c. Multiplying every score by 2 will multiply the mean, the standard deviation, and your score. However, your z-score (your position within the distribution) will not change.

4. a. X = 48 corresponds to z = 0. In the new distribution this location corresponds to X = 100.
   b. X = 50 corresponds to z = +0.25 which corresponds to X = 105 in the new distribution.
   c. X = 44 corresponds to z = –0.50. X = 90 in the new distribution.
   d. X = 32 corresponds to z = –2.00. X = 60 in the new distribution.

CHAPTER 6

PROBABILITY

CHAPTER SUMMARY

The goals of Chapter 6 are:
1. To introduce the concept of probability.
2. To demonstrate how the unit normal table can be used to find the proportions of a normal distribution corresponding to any specific z-score location.
3. To demonstrate how the combination of z-scores and the unit normal table can be used to find probabilities any score (X value) from a normal distribution, and how the table can be used to find the score corresponding to any specific proportion.
4. To demonstrate how the normal distribution can be used to find binomial probabilities.
5. To demonstrate how probability will be used as part of the inferential statistics in future chapters.

Probability

In this chapter we introduce the concept of probability as a method for measuring and quantifying the likelihood of obtaining a specific sample from a specific population. We define probability as a fraction or a proportion. In particular, the probability of any specific outcome is determined by a ratio comparing the frequency of occurrence for that outcome relative to the total number of possible outcomes.

Whenever the scores in a population are variable it is impossible to predict with perfect accuracy exactly which score or scores will be obtained when you take a sample from the population. In this situation researchers rely on probability to determine the relative likelihood for specific samples. Thus, although a researcher may not be able to predict exactly which value(s) will be obtained for a sample, it is possible to determine exactly which outcomes have high probability and which have low probability.

Probability is determined by a fraction or proportion. When a population of scores is represented by a frequency distribution, probabilities can be defined by proportions of the distribution. In graphs, probability can be defined as a proportion of area under the curve.

Probability and the Normal Distribution
If a vertical line is drawn through a normal distribution, several things occur.
1. The exact location of the line can be specified by a z-score.
2. The line divides the distribution into two sections. The larger section is called the **body** and the smaller section is called the **tail**.

The **unit normal table** lists several different proportions corresponding to each z-score location. Column A of the table lists z-score values. For each z-score location, columns B and C list the proportions in the body and tail, respectively. Finally, column D lists the proportion between the mean and the z-score location. Because probability is equivalent to proportion, the table values can also be used to determine probabilities.

To find the probability corresponding to a particular score (X value), you first transform the score into a z-score, then look up the z-score in the table and read across the row to find the appropriate proportion/probability. To find the score (X value) corresponding to a particular proportion, you first look up the proportion in the table, read across the row to find the corresponding z-score, and then transform the z-score into an X value.

Percentiles and Percentile Ranks

In Box 6.1 we noted that the **percentile rank** for a specific X value is the percentage of individuals with scores at or below that value. When a score is referred to by its rank, the score is called a **percentile**. The percentile rank for a score in a normal distribution is simply the proportion to the left of the score.

Probability and the Binomial Distribution

Binomial distributions are formed by a series of observations (for example, 100 coin tosses) for which there are exactly two possible outcomes (heads and tails). The two outcomes are identified as A and B, with probabilities of p(A) = p and p(B) = q. The distribution shows the probability for each value of X, where X is the number of occurrences of A in a series of n observations. When pn and qn are both greater than 10, the binomial distribution is closely approximated by a normal distribution with a mean of μ = pn and a standard deviation of

$$\sigma = \sqrt{npq}.$$

In this situation, a z-score can be computed for each value of X and the unit normal table can be used to determine probabilities for specific outcomes.

Probability and Inferential Statistics

Probability is important because it establishes a link between samples and populations. For any known population it is possible to determine the probability of obtaining any specific sample. In later chapters we will use this link as the foundation for inferential statistics. The general goal of inferential statistics is to use the information from a sample to reach a general conclusion (inference) about an unknown population.

Typically a researcher begins with a sample. If the sample has a high probability of being obtained from a specific population, then the researcher can conclude that the sample is likely to have come from that population. On the other hand, if the sample has a very low probability of being obtained from a specific population, then it is reasonable for the researcher to conclude that the specific population is probably not the source for the sample.

LEARNING OBJECTIVES

1. Know how to determine the probability of an event.

2. Be able to use the unit normal table to determine the probabilities for events that are normally distributed.

3. Be able to combine z-score calculations (Chapter 5) and the unit normal table to find probabilities for scores from a normal distribution, or to find the scores associated with specific proportions.

4. Be able to determine binomial probabilities using the normal approximation.

NEW TERMS AND CONCEPTS

The following terms were introduced in this chapter. You should be able to define or describe each term and, where appropriate, describe how each term is related to other terms in the list.

Probability	Probability is defined as a proportion, a specific part out of the whole set of possibilities.
Proportion	A part of the whole usually expressed as a fraction.
Random sample	A sample obtained using a process that gives every individual an equal chance of being selected and keeps the probability of being selected constant over a series of selections.
Sampling with replacement	A sampling technique that returns the current selection to the population before the next selection is made. A required part of random sampling.
Normal distribution	A symmetrical, bell-shaped distribution with proportions corresponding to those listed in the unit normal table.
Unit normal table	A table listing proportions corresponding to each z-score location in a normal distribution.
Percentile	A score that is identified by the percentage of the distribution that falls below its value.
Percentile rank	The percentage of a distribution that falls below a specific score.
Binomial distribution	The distribution of probabilities, for each possible outcome, for a series of observations of a dichotomous variable.

NEW FORMULAS

$$p(A) = \frac{\text{Number of ways event A can occur}}{\text{Total number of possible outcomes}}$$

$$z = \frac{X - pn}{\sqrt{npq}} \qquad \mu = pn \qquad \sigma = \sqrt{npq}$$

STEP BY STEP

<u>Finding the probability associated with a specified score</u>. The general process involves converting the score (X) into a z-score, then using the unit normal table to find the probability associated with the z-score. To demonstrate this process, we will find the probability of randomly selecting a score greater than 95 from a normal distribution with $\mu = 100$ and $\sigma = 10$.

Step 1: Sketch the distribution and identify the mean and standard deviation. Then, find the approximate location of the specified score and draw a vertical line through the distribution. For this example, X = 95 is located below the mean by roughly one-half of the standard deviation.

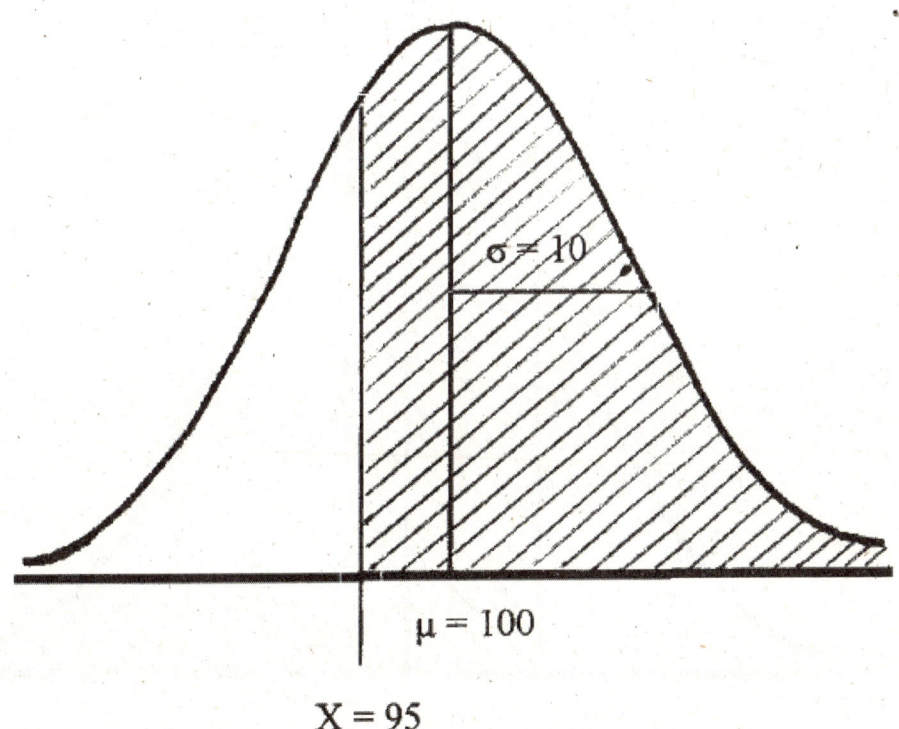

PROBABILITY

Step 2: Read the problem again to determine whether you want the proportion greater than the score (right of your line) or less than the score (left of the line). Then shade in the appropriate portion of the distribution. For this example we want the proportion consisting of scores greater than 95, so shade in the portion to the right of X = 95.

Step 3: Look at your sketch and make an estimate of the proportion that has been shaded. Remember, the mean divides the distribution in half with 50% on each side. For this example, we have shaded more than 50% of the distribution. The shaded area appears to be about 60% or 70% of the distribution.

Step 4: Transform the X value into a z-score. For this example, X = 95 corresponds to z = –0.50.

Step 5: Look up the z-score value in the unit normal table. (Ignore the + or – sign.) Find the proportions in the table that correspond to your z-score. Usually, the proportions in the tail (column C) and in the body (column B) will be sufficient to find the answer. Write these two proportions in the appropriate places on your figure.

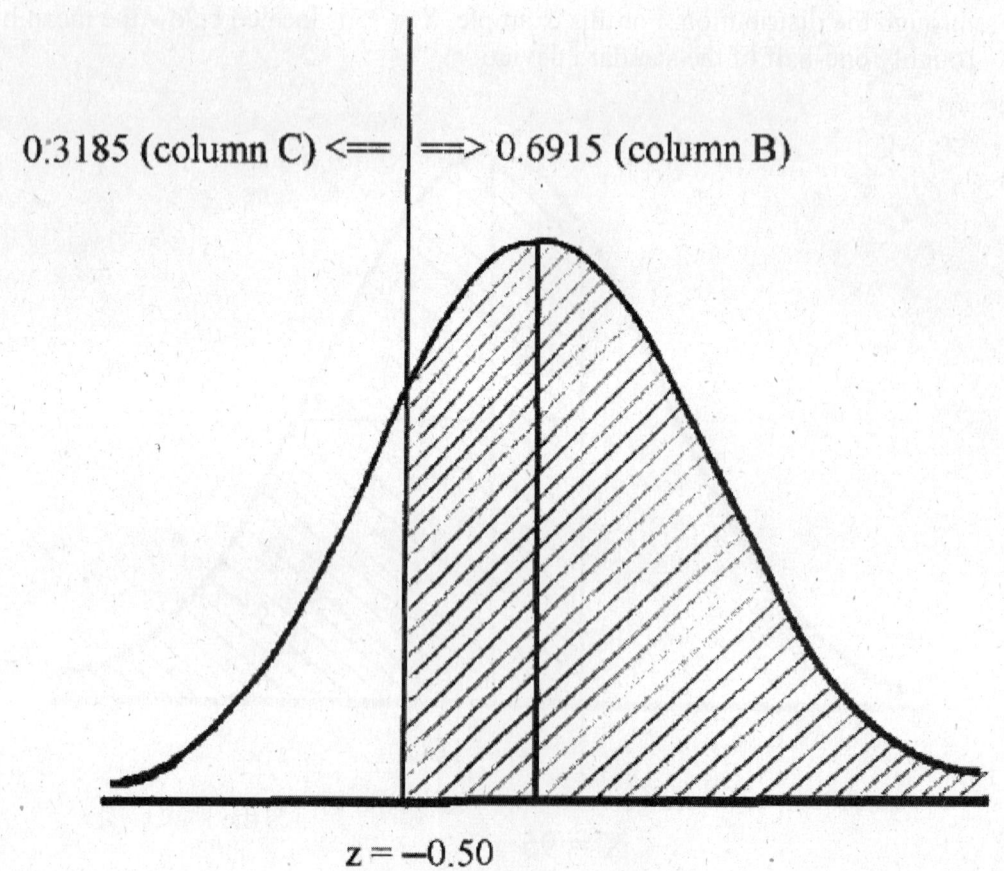

Step 6: You must use the proportions from the table to find the value that corresponds to the shaded area of your figure. For this example, the column B proportion, 0.6915, corresponds to the shaded area.

Step 7: Compare your final answer with the estimate you made in Step 4. If your answer does not agree with your preliminary estimate, re-work the problem.

HINTS AND CAUTIONS

1. In using probability, you should be comfortable in converting fractions into decimals or percentages. These values all represent ways of expressing portions of the whole. If you have difficulty with fractions or decimals, review the section on proportions in the math review appendix of your textbook.

2. It usually helps to restate a probability problem as a question about proportion. For example, the problem, "What is the probability of selecting an ace from a deck of cards?" becomes, "What proportion of the deck is composed of aces?"

3. When using the unit normal table to answer probability questions, you should always start by sketching a normal distribution and shading in the area of the distribution for which you need a proportion.

4. When you are computing probabilities for a binomial distribution, remember to use the real limits for each score. For example, X = 18 actually corresponds to an interval from 17.5 to 18.5. When you are looking for scores greater than 18, you should begin at the upper real limit of 18.5.

SELF TEST

True/False Questions

1. The value for a probability can never be less than zero or greater than 1.00.

2. A jar contains 10 red marbles and 20 blue marbles. If you take a *random sample* of two marbles from this jar and the first marble is blue, then the probability that the second marble is blue is p = 19/29.

3. For a normal distribution, proportions in the right-hand tail are positive and proportions in the left-hand tail are negative.

4. For a normal distribution, the z-score boundary that separates the lowest 2.5% of the scores from the rest is z = –1.96.

5. For a normal distribution, the proportion in the tail beyond z = 1.50 is p = 0.0668.

6. For a normal distribution, the proportion located between z = −1.00 and z = +1.00 is p = 34.13%

7. For a population with a mean of μ = 80 and σ = 10, the score that separates the lowest 25% of the scores from the rest is X = 86.7.

8. For a population with a mean of μ = 80 and σ = 10, only 2.28% of the scores are greater than X = 100.

9. In a binomial situation, p + q = 1.00.

10. The binomial distribution for p = q = ½ and n = 64 has a standard deviation of σ = 16.

Multiple-Choice Questions

1. If you are randomly selecting one student from a class of 7 males and 23 females, what is the probability of selecting a male?
    a. 7/23
    b. 7/30
    c. 23/30
    d. 23/7

2. A jar contains 40 red marbles, 20 blue marbles, and 10 black marbles. If you take a *random sample* of n = 3 marbles from this jar, and the first two marbles are both red, what is the probability that the third marble will be black?
    a. 1/68
    b. 1/70
    c. 10/68
    d. 10/70

3. The proportion of a normal distribution that corresponds to values greater than z = 1.00 is p = 0.1587. What is the proportion that corresponds to values less than z = −1.00?
    a. 0.8413
    b. −0.8413
    c. 0.1587
    d. −0.1587

4. For a normal distribution, what z-score value separates the lowest 10% of the scores from the rest of the distribution?
    a. z = 0.10
    b. z = 0.90
    c. z = 1.28
    d. z = −1.28

5. What percentage of a normal distribution is located in the tail beyond z = 2.00?
    a. 1.14%
    b. 2%
    c. 2.28%
    d. 97.72%

6. If the tail of a normal distribution contains exactly 2.5% of the scores, then what is the z-score value that separates the tail from the body of the distribution?
    a. z = 1.96
    b. z = –1.96
    c. either z = 1.96 or z = –1.96
    d. none of the above is correct

7. For a normal distribution with μ = 90 and σ = 5, what proportion of the scores have values greater than X = 87?
    a. 0.2743
    b. 0.7257
    c. 0.1151
    d. 0.8849

8. For a normal distribution with μ = 50 and σ = 10, what is the probability of randomly selecting a score greater than X = 56?
    a. 0.3446
    b. 0.6554
    c. 0.2743
    d. 0.7257

9. For a normal distribution with μ = 100 and σ = 5, what is the probability of selecting a score between X = 90 and X = 110? In symbols, what is p(110 < X < 90)?
    a. 0.4772
    b. 0.9772
    c. 0.9544
    d. 0.1336

10. For any normal distribution, what z-score separates the highest 80% of the scores from the rest of the distribution?
    a. z = 0.84
    b. z = –0.84
    c. z = 1.28
    d. z = –1.28

11. For any normal distribution, what are the z-score values that form the boundaries for the middle 80%?
    a. z = ±0.25
    b. z = ±0.52
    c. z = ±0.84
    d. z = ±1.28

12. Scores on the SAT form a normal distribution with $\mu = 500$ and $\sigma = 100$. What is the minimum SAT score needed to be in the top 60% of the distribution?
    a. X = 475
    b. X = 525
    c. X = 372
    d. X = 628

13. For a binomial distribution with p = 1/3 and q = 2/3, how large a sample is needed to justify using the normal approximation?
    a. 10
    b. 15
    c. 30
    d. 60

14. A binomial distribution has p = 0.2, q = 0.8, and n = 100. The normal approximation for this distribution will have a standard deviation of _____.
    a. 4
    b. 16
    c. 20
    d. 80

15. A binomial distribution has p = q = ½ and n = 100. Using the normal approximation to this distribution, the probability of selecting a score greater than X = 60 corresponds to _____?
    a. p(z > 9.5/5)
    b. p(z > 10/5)
    c. p(z > 10.5/5)
    d. p(z > 11/5)

Other Questions

1. Assume a normal distribution for each question.
    a. What proportion of the distribution consists of z-scores greater than 0.25?
    b. What is the probability of obtaining a z-score less than 0.50?
    c. What is the probability of obtaining a z-score greater than –1.50?
    d. What is the probability of obtaining a z-score between +1.00 and –1.00?

2. Find the z-score that separates a normal distribution into the following two portions:
   a. separate the lowest 80% from the highest 20%
   b. separate the lowest 90% from the highest 10%
   c. separate the lowest 15% from the highest 85%
   d. separate the lowest 40% from the highest 60%

3. Find the following probabilities for a normal distribution with $\mu = 80$ and $\sigma = 12$.
   a. $p(X > 86)$
   b. $p(X > 77)$
   c. $p(X < 95)$
   d. $p(X < 68)$

4. For a normal distribution with $\mu = 80$ and $\sigma = 12$, find the X value associated with each of the following proportions:
   a. What X value separates the distribution into the top 40% versus the bottom 60%?
   b. What is the minimum X value needed to be in the top 25% of the distribution?
   c. What X value separates the top 60% from the bottom 40% of the distribution?

5. A psychologist is testing people's sense of direction by leading them through a maze in the basement of the psychology building and then asking them to identify which of four directions is north. In a sample of 48 people, 17 correctly identified the direction. What is the probability that 17 or more people would correctly identify the direction if they were simply guessing?

ANSWERS TO SELF TEST

True/False Answers

1. True
2. False. Random sampling requires replacement.
3. False. Proportions are always positive, on either side of the distribution.
4. True
5. True
6. False. The total proportion is two times 34.13%.
7. False. $X = 86.7$ separates the highest 25% from the rest.
8. True
9. True
10. False. The standard deviation is the square root of 16.

Multiple-Choice Answers

1. b There are 7 males out of a total of 30 students.
2. d Remember, a *random sample* includes sampling with replacement.

3. c The normal distribution is symmetrical.
4. d With 0.1000 in the left-hand tail, the z-score is z = –1.28.
5. c The proportion in the tail is 0.0228.
6. c The 2.5% could be in either tail.
7. b For z = –0.60, the proportion in the body is 0.7257.
8. c For z = 0.60, the proportion in the tail is 0.2743.
9. c The proportion is 0.4772 on each side of the mean.
10. b The boundary for the top 80% is z = –0.84.
11. d There is 10% in each tail.
12. a The top 60% corresponds to z = –0.25 which corresponds to X = 475.
13. c Both pn and qn must be at least 10.
14. a The standard deviation is the square root of 16.
15. c X = 60 corresponds to an interval from 59.5 to 60.5. To be greater than 60 the score must be greater than 60.5.

Other Answers

1. a. p(z > 0.25) = 0.4013
   b. p(z < 0.50) = 0.6915
   c. p(z > –1.50) = 0.9332
   d. p(–1.00 < z > 1.00) = 0.6826

2. a. z = 0.84   c. z = –1.04
   b. z = 1.28   d. z = –0.25

3. a. z = 0.50 and p = 0.3085
   b. z = –0.25 and p = 0.5987
   c. z = 1.25 and p = 0.8944
   d. z = –1.00 and p = 0.1587

4. a. z = 0.25 and X = 83
   b. z = 0.67 and X = 88.04
   c. z = –0.25 and X = 77

5. The probability of guessing correctly is p = ¼. The normal approximation to the binomial distribution has μ = 12 and σ = 3. A score of X = 17 or more corresponds to a real limit of X = 16.5. The z-score is 4.5/3 = 1.50 and p = 0.0668.

CHAPTER 7

THE DISTRIBUTION OF SAMPLE MEANS

CHAPTER SUMMARY

The goals for Chapter 7 are:
1. To introduce the distribution of sample means and identify its parameters.
2. To demonstrate how z-scores can be used to identify the location of a specific sample mean within the distribution of sample means.
3. To demonstrate how the unit normal table can be used to find probabilities for sample means.
4. To introduce the concept of standard error as a critical component of inferential statistics.

The Distribution of Sample Means
 In the previous two chapters we presented the statistical procedures for computing z-scores and finding probabilities associated with individual scores, X-values. In order to find z-scores or probabilities, the first requirement is that you must know about *all the possible X values*, that is, the entire distribution. A z-score tells where an individual X is located relative to all the other X values in the distribution. To find a probability, we simply identified a proportion of all the possible X values.
 In Chapter 7 we extend the concepts of z-scores and probability to samples of more than one score. Specifically, we will compute z-scores and find probabilities for sample means. To accomplish this task, the first requirement is that you must know about *all the possible sample means*, that is, the entire distribution of Ms. Once this distribution is identified, then

1. A z-score can be computed for each sample mean. The z-score tells where the specific sample mean is located relative to all the other sample means.
2. The probability associated with a specific sample mean can be defined as a proportion of all the possible sample means.

The **distribution of sample means** is defined as the set of means from all the possible random samples of a specific size (n) selected from a specific population. This distribution has well-defined (and predictable) characteristics that are specified in the Central Limit Theorem:
1. The mean of the distribution of sample means is called the **Expected Value of M** and is always equal to the population mean µ.

2. The standard deviation of the distribution of sample means is called the **Standard Error of M** and is computed by

$$\sigma_M = \frac{\sigma}{\sqrt{n}} \quad \text{or} \quad \sigma_M = \sqrt{\frac{\sigma^2}{n}}$$

3. The shape of the distribution of sample means tends to be normal. It is guaranteed to be normal if either a) the population from which the samples are obtained is normal, or b) the sample size is n = 30 or more.

The concept of the distribution of sample means and its characteristics should be intuitively reasonable. First, you should realize that sample means are variable. If two (or more) samples are selected from the same population, the two samples probably will have different means. Second, although the samples will have different means, you should expect the sample means to be close to the population mean. That is, the sample means should "pile up" around μ. Thus, the distribution of sample means tends to form a normal shape with an expected value of μ. Finally, you should realize that an individual sample mean probably will not be identical to its population mean; that is, there will be some "error" between M and μ. Some sample means will be relatively close to μ and others will be relatively far away. The standard error provides a measure of the standard distance between M and μ.

z-Scores and Location within the Distribution of Sample Means
Within the distribution of sample means, the location of each sample mean can be specified by a z-score,

$$z = \frac{M - \mu}{\sigma_M}$$

As always, a positive z-score indicates a sample mean that is greater than μ and a negative z-score corresponds to a sample mean that is smaller than μ. The numerical value of the z-score indicates the distance between M and μ measured in terms of the standard error.

Probability and Sample Means
Because the distribution of sample means tends to be normal, the z-score value obtained for a sample mean can be used with the unit normal table to obtain probabilities. The procedures for computing z-scores and finding probabilities for sample means are essentially the same as we used for individual scores (in Chapters 5 and 6). However, when you are using sample means, you must remember to consider the sample size (n) and compute the standard error (σ_M) before you start any other computations. Also, you must be sure that the distribution of sample means satisfies at least one of the criteria for normal shape before you can use the unit normal table.

The Standard Error of M

Standard error is perhaps the single most important concept in inferential statistics. The standard error of M is defined as the standard deviation of the distribution of sample means and measures the standard distance between a sample mean and the population mean. Thus, the Standard Error of M provides a measure of how accurately, on average, a sample mean represents its corresponding population mean.

The magnitude of the standard error is determined by two factors: σ and n. The population standard deviation, σ, measures the standard distance between a single score (X) and the population mean. Thus, the standard deviation provides a measure of the "error" that is expected for the smallest possible sample, when n = 1. As the sample size is increased, it is reasonable to expect that the error should decrease. In simple terms, the larger the sample, the more accurately it should represent its population. The formula for standard error reflects the intuitive relationship between standard deviation, sample size, and "error."

$$\sigma_M = \frac{\sigma}{\sqrt{n}}$$

As the sample size increases, the error decreases. As the sample size decreases, the error increases. At the extreme, when n = 1, the error is equal to the standard deviation.

LEARNING OBJECTIVES

1. For any specific sampling situation, you should be able to define and describe the distribution of sample means by identifying its shape, the expected value of M, and the standard error of M.

2. You should be able to define and calculate the standard error of M.

3. You should be able to compute a z-score that specifies the location of a particular sample mean within the distribution of sample means.

4. Using the distribution of sample means, you should be able to compute the probability of obtaining specific values for a sample mean obtained from a given population.

5. You should be able to incorporate a visual presentation of standard error into a graph presenting means for a set of different samples. In addition, you should be able to use the visual presentation of standard error to help determine whether the obtained difference between two sample means reflects a "real" difference in the populations, or whether the sample mean difference is simply due to chance.

NEW TERMS AND CONCEPTS

The following terms were introduced in this chapter. You should be able to define or describe each term and, where appropriate, describe how each term is related to other terms in the list.

Distribution of sample means — The set of sample means from all the possible random samples for a specific sample size (n) from a specific population.

Sampling distribution — A distribution of statistics (as opposed to a distribution of scores). The distribution of sample means is an example of a sampling distribution.

Expected value of M — The mean of the distribution of sample means. The average of the M values.

Standard error of M — The standard deviation of the distribution of sample means. The standard distance between a sample mean and the population mean.

The central limit theorem — A mathematical theorem that specifies the characteristics of the distribution of sample means.

NEW FORMULAS

$$\sigma_M = \frac{\sigma}{\sqrt{n}} \quad \text{or} \quad \sigma_M = \sqrt{\frac{\sigma^2}{n}}$$

$$z = \frac{M - \mu}{\sigma_M}$$

STEP BY STEP

<u>Computing Probabilities for Sample Means</u>: You should recall that we have defined probability as being equivalent to proportion. Thus, the probability associated with a specific sample mean can be defined as a specific proportion of the distribution of sample means. Because the distribution of sample means tends to be normal, you can use

z-scores and the unit normal table to determine proportions or probabilities. The following example demonstrates the details of this process.

For a normal population with $\mu = 60$ and $\sigma = 12$, what is the probability of selecting a random sample of n = 36 scores with a sample mean greater than 64? In symbols, p(M > 64) = ?

Step 1: Rephrase the probability question as a proportion question. For this example, "Out of all the possible sample means for n = 36, what proportion have values greater than 64?"

Step 2: We are looking for a specific proportion of "all the possible sample means." The set of "all possible sample means" is the distribution of sample means. Therefore, the next step is to sketch the distribution. Show the expected value and standard error in your sketch. Caution: Be sure to use the standard error, not the standard deviation.
For this example, the distribution of sample means will have an expected value of $\mu = 60$, a standard error of $\sigma_M = 12/\sqrt{36} = 2$, and it will be a normal distribution because the original population is normal (also because n > 30).
Caution: If the distribution of sample means is not normal, you cannot use the unit normal table to find probabilities.

Step 3: Find the approximate location of the specified sample mean and draw a vertical line through the distribution. For this example, M = 64 is located above the mean by roughly two times the standard deviation.

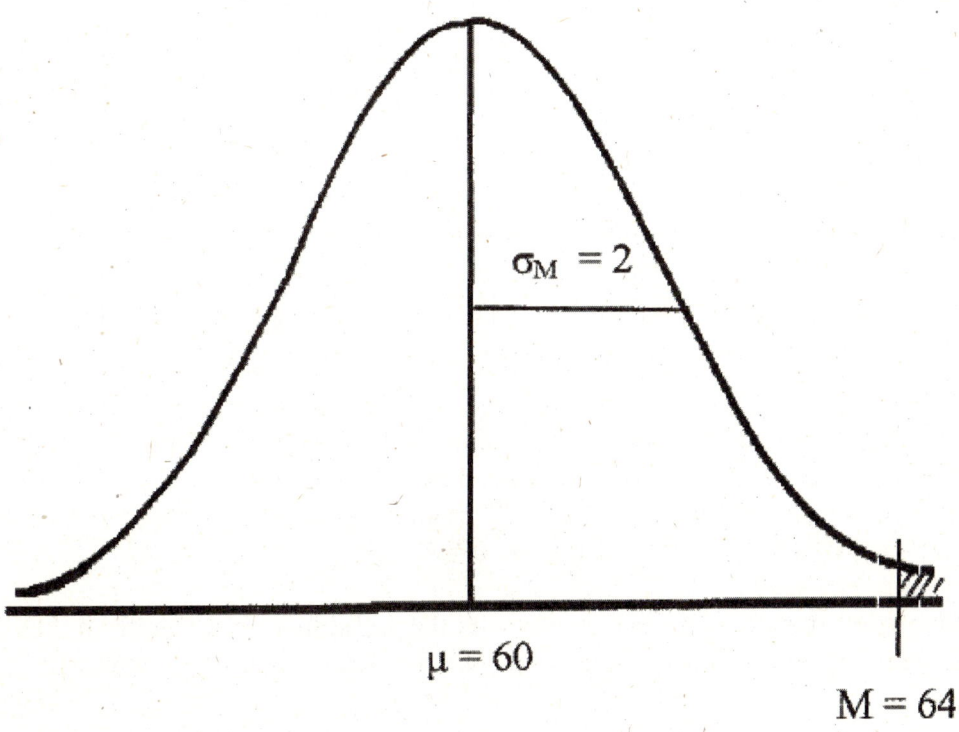

THE DISTRIBUTION OF SAMPLE MEANS

Step 4: Determine whether the problem asked for the proportion greater than or less than the specific M. Then shade in the appropriate area in your sketch. For this example, we want the area greater than M = 64 so shade in the area on the right-hand side of the line.

Step 5: Look at your sketch and make a preliminary estimate of the proportion that is shaded. For this example, we have shaded a very small part of the whole distribution, probably 5% or less.

Step 6: Compute the z-score for the specified sample mean. Be sure to use the z-score formula for sample means. For this example, M = 64 corresponds to z = +2.00.

$$z = \frac{M - \mu}{\sigma_M} = \frac{64 - 60}{2} = \frac{4}{2} = 2.00$$

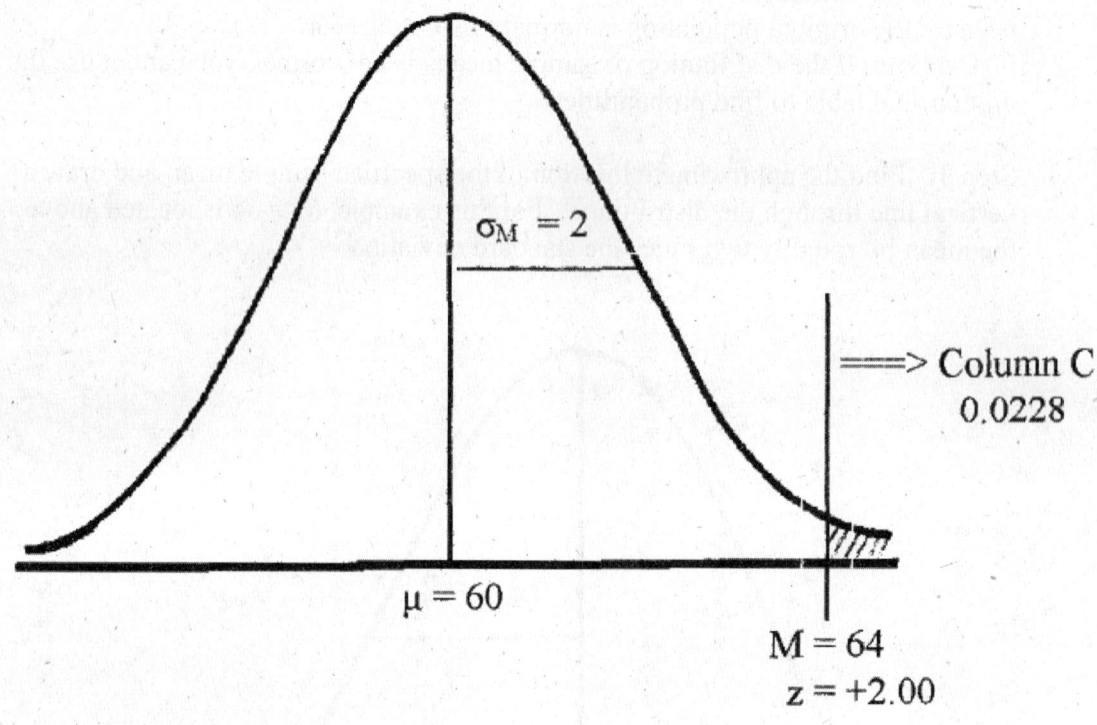

Step 7: Look up the z-score in the unit normal table and find the two proportions in columns B and C. For this example, the value in column C (the tail of the distribution) corresponds exactly to the proportion we want.
    p(M > 64) = p(z > +2.00) = .0228

Step 8: Compare your final answer with the preliminary estimate from Step 5. Be sure that your answer is in agreement with the estimate you made earlier.

HINTS AND CAUTIONS

1. Whenever you encounter a question about a sample mean you must remember to use the distribution of sample means and not the original population distribution. The population distribution contains scores (not sample means) and therefore should be used only when you have a question about an individual score (n = 1).

2. The key to working with the distribution of sample means is the standard error of M:

$$\sigma_M = \frac{\sigma}{\sqrt{n}}$$

Remember, larger samples tend to be more accurate (less error) than smaller samples: The larger the sample, the smaller the difference between the sample mean and the population mean. The sample size, n, is a crucial factor in determining the standard error between a sample and its population.

SELF TEST

True/False Questions

1. A sample of n = 25 scores is selected from a population with a mean of μ = 80 and a standard deviation of σ = 20. The expected value for the sample mean is 80.

2. A sample of n = 25 scores is selected from a population with a mean of μ = 80 and a standard deviation of σ = 20. The standard error for the sample mean is 20.

3. Two samples probably will have different means even if they are both the same size and they are both selected from the same population.

4. As the sample size increase, the standard error also increases.

5. The mean for a sample of n = 4 scores has a standard error σ = 5 points. This sample was selected from a population with a standard deviation of σ = 20.

6. The mean for a sample of n = 16 scores has an expected value of 50. This sample was selected from a population with a mean of μ = 50.

7. If a sample consists of at least n = 30 scores, then the sample mean will be equal to the population mean.

8. The standard error for a sample mean can never be greater than the standard deviation of the population from which the sample is selected.

9. A researcher obtained M = 27 for a sample of n = 36 scores selected from a population with μ = 30 and σ = 18. This sample mean corresponds to a z-score of z = –1.00.

10. A sample of n = 9 scores is selected from a normal population with a mean of μ = 80 and a standard deviation of σ = 12. The probability that the sample mean will be greater than M = 86 is equal to the probability of obtaining a z-score greater than z = 1.50

Multiple-Choice Questions

1. For a population with μ = 80 and σ = 20, the distribution of sample means based on n = 16 will have an expected value of _____ and a standard error of _____.
    a. 5, 80
    b. 80, 5
    c. 20, 20
    d. 80, 1.25

2. The distribution of sample means _____.
    a. is always a normal distribution.
    b. will be normal *only if* the population distribution is normal.
    c. will be normal *only if* the sample size is at least n = 30.
    d. we be normal *if either* the population is normal or the sample size is n ≥ 30.

3. If other factors are held constant, the expected value of the sample mean will _____ as the sample size increases.
    a. increase
    b. decrease
    c. stay constant

4. If other factors are held constant, the standard error of M will _____ as the sample size increases.
    a. increase
    b. decrease
    c. stay constant

5. A sample of n = 100 scores is selected from a population with μ = 80 with σ = 20. On average, how much error is expected between the sample mean and the population mean?
    a. 0.2 points
    b. 0.8 points
    c. 2 points
    d. 4 points

6. A sample of n = 9 scores is determined to have a standard error of 4 points. What is the standard deviation for the population from which the sample was obtained?
   a. 4/3
   b. 4
   c. 12
   d. 36

7. A random sample is selected from a population with a standard deviation of σ = 20. If the sample mean has a standard error of 2 points, how many scores are in the sample?
   a. n = 5
   b. n = 10
   c. n = 25
   d. n = 100

8. Which of the following samples would have the smallest standard error?
   a. n = 25 scores from a population with σ = 10
   b. n = 25 scores from a population with σ = 20
   c. n = 100 scores from a population with σ = 10
   d. n = 100 scores from a population with σ = 20

9. A normal population has μ = 50 and σ = 8. A random sample of n = 4 scores from this population has a mean of 54. What is the z-score for this sample mean?
   a. +0.50
   b. +1.00
   c. +2.00
   d. +4.00

10. A sample of n = 4 scores has a standard error of 10 points. For the same population, what is the standard error for a sample of n = 16 scores?
    a. 1
    b. 2.5
    c. 5
    d. 10

11. A random sample of n = 4 scores is obtained from a population with a mean of μ = 80 and a standard deviation of σ = 10. If the sample mean is M = 90, what is the z-score for the sample mean?
    a. z = 20.00
    b. z = 5.00
    c. z = 2.00
    d. z = 1.00

12. For a normal population with μ = 80 and σ = 20 which of the following sample means is an extreme and unrepresentative value?
   a. M greater than 90 for a sample of n = 4
   b. M greater than 85 for a sample of n = 4
   c. M greater than 90 for a sample of n = 25
   d. M greater than 85 for a sample of n = 25

13. A sample of n = 4 scores is selected from a population with μ = 50 and σ = 12. If the sample mean is M = 56, what is the z-score for this sample mean?
   a. 0.50
   b. 1.00
   c. 2.00
   d. 4.00

14. For a normal population with a mean of μ = 80 and a standard deviation of σ = 10, what is the probability of obtaining a sample mean greater than M = 75 for a sample of n = 25 scores?
   a. p = 0.0062
   b. p = 0.9938
   c. p = 0.3085
   d. p = 0.6915

15. A random sample of n = 4 scores is selected from a normally distributed population with μ = 80 and σ = 12. What is the probability that the sample mean will be greater than 86?
   a. 0.0228
   b. 0.0668
   c. 0.1587
   d. 0.3085

Other Questions

1. A population has a mean of μ = 100 and a standard deviation of σ = 10.
   a. If a single score is randomly selected from this population, how much distance, on average, should you find between the score and the population mean?
   b. If a sample of n = 4 scores is randomly selected from this population, how close on the average should the sample mean be to the population mean?
   c. If a sample of n = 100 scores is randomly selected from this population, how close on the average should the sample mean be to the population mean?

2. Each of the following samples was obtained from a population with μ = 100 and σ = 10. Find the z-score for each sample mean.
   a. M = 90 for a sample of n = 4
   b. M = 90 for a sample of n = 25
   c. M = 102 for a sample of n = 4
   d. M = 102 for a sample of n = 100

3. For a normal population with μ = 70 and σ = 12:
   a. What is the probability of obtaining a sample mean greater than 73 for a sample of n = 36 scores?
   b. What is the probability of obtaining a sample mean greater than 73 for a sample of n = 9 scores?

4. Given a normal population with μ = 40 and σ = 8, what is the probability of obtaining a sample mean between 39 and 41 for a sample of n = 16 scores?

ANSWERS TO SELF TEST

True/False Answers

1. True
2. False. The standard error is 20 divided by the square root of 25.
3. True
4. False. As sample size increases, the standard error decreases.
5. False. The population standard deviation is equal to 10.
6. True
7. False. With a sample size of n = 30 or more, the sample mean should be close to μ but probably not equal to μ.
8. True
9. True
10. True

Multiple-Choice Answers

1. b The expected value is μ and the standard error is $\sigma/\sqrt{n}$.
2. d Either criterion is sufficient for a normal distribution.
3. c The expected value of M is equal to μ, independent of n.
4. b The standard error is inversely related to the sample size.
5. c The standard error is 2 points.
6. c The standard error is 12/3 = 4.
7. d The standard error is 20/10 = 2. $10 = \sqrt{100}$
8. c The smallest standard deviation and the biggest sample will produce the smallest error.
9. b With a mean of 50 and a standard error of 4, M = 54 corresponds to z = 1.00.
10. c The population standard deviation is 20.
11. c M = 90 is above the mean by 2 standard errors.
12. c With n = 25, M = 90 corresponds to z = 2.50.
13. b M = 56 is above the mean by exactly one standard error.
14. a M = 75 corresponds to z = –2.50.
15. c With a mean of 80 and a standard error of 6, M = 86 corresponds to z = 1.00 which has 0.1587 in the tail.

Other Answers

1. a. The standard deviation, $\sigma = 10$, measures the standard distance between a score and the population mean.
   b. For n = 4, the standard error is $10/\sqrt{4} = 5$ points.
   c. For a sample of n = 100, the standard error is 1 point.

2. a. The standard error is 5, and z = –2.00.
   b. The standard error is 2, and z = –5.00.
   c. The standard error is 5, and z = +0.40.
   d. The standard error is 1, and z = +2.00.

3. a. The standard error is 2. p(M > 73) = p(z > +1.50) = 0.0668.
   b. The standard error is 4. p(M > 73) = p(z < 0.75) = 0.2266.

4. The standard error is 2. The probability is p(–0.50 < z < +0.50) = 0.3830. [2(.1915) using the value from Column D in the unit normal table]

CHAPTER 8

INTRODUCTION TO HYPOTHESIS TESTING

CHAPTER SUMMARY

The goals of Chapter 8 are:
1. To introduce the statistical technique of hypothesis testing.
2. To introduce the concepts of null hypothesis, alpha level, and critical region.
3. To introduce the types of errors that can occur in a hypothesis test.
4. To demonstrate the differences between one-tailed tests and two-tailed tests.
5. To introduce the concept of effect size as a supplement to a hypothesis test.
6. To introduce the concept of power for a hypothesis test.

Hypothesis Testing
   The general goal of a hypothesis test is to rule out chance (sampling error) as a plausible explanation for the results from a research study. In this chapter, we introduced hypothesis testing as a technique to help determine whether a specific treatment has an effect on the individuals in a population. The hypothesis test is used to evaluate the results from a research study in which
   1. A sample is selected from the population.
   2. The treatment is administered to the sample.
   3. After treatment, the individuals in the sample are measured.
If the individuals in the sample are noticeably different from the individuals in the original population, we have evidence that the treatment has an effect. However, it is also possible that the difference between the sample and the population is simply sampling error (see Figure 1.2 in your textbook). The purpose of the hypothesis test is to decide between two explanations:
   1. The difference between the sample and the population can be explained by sampling error (there does not appear to be a treatment effect)
   2. The difference between the sample and the population is too large to be explained by sampling error (there does appear to be a treatment effect).

The Null Hypothesis, the Alpha Level, the Critical Region, and the Test Statistic
   The following four steps outline the process of hypothesis testing and introduce some of the new terminology.

   1. State the hypotheses and select an α level. The **null hypothesis**, H_0, always states that the treatment has no effect (no change, no difference). According to the null hypothesis, the population mean after treatment is the same is it was before treatment. The **α level** establishes a criterion, or "cut-off", for making a decision about the null hypothesis. The alpha level also determines the risk of a Type I error.

2. Locate the critical region. The **critical region** consists of outcomes that are very unlikely to occur if the null hypothesis is true. That is, the critical region is defined by sample means that are almost impossible to obtain if the treatment has no effect. The phrase "almost impossible" means that these samples have a probability (p) that is less than the alpha level.

3. Compute the test statistic. The **test statistic** (in this chapter a z-score) forms a ratio comparing the obtained difference between the sample mean and the hypothesized population mean versus the amount of difference we would expect without any treatment effect (the standard error).

4. A large value for the test statistic shows that the obtained mean difference is more than would be expected if there is no treatment effect. If it is large enough to be in the critical region, we conclude that the difference is **significant** or that the treatment has a significant effect. In this case we reject the null hypothesis. If the mean difference is relatively small, then the test statistic will have a low value. In this case, we conclude that the evidence from the sample is not sufficient, and the decision is fail to reject the null hypothesis.

Errors in Hypothesis Tests

Conclusions from hypothesis tests can get a bit tricky. Just because the sample mean (following treatment) is different from the original population mean does not necessarily indicate that the treatment has caused a change. You should recall that there usually is some discrepancy between a sample mean and the population mean simply as a result of sampling error. This point was first demonstrated in Figure 1.2. Because the hypothesis test relies on sample data, and because sample data are not completely reliable, there is always the risk that misleading data will cause the hypothesis test to reach a wrong conclusion. Two types of error are possible:

1. A **Type I error** occurs when the sample data appear to show a treatment effect when, in fact, there is none. In this case the researcher will reject the null hypothesis and falsely conclude that the treatment has an effect. Type I errors are caused by unusual, unrepresentative samples. Just by chance the researcher selects an extreme sample with the result that the sample falls in the critical region even though the treatment has no effect. Fortunately, the hypothesis test is structured so that Type I errors are very unlikely; specifically, the probability of a Type I error is equal to the alpha level.

2. A **Type II error** occurs when the sample does not appear to have been affected by the treatment when, in fact, the treatment does have an effect. In this case, the researcher will fail to reject the null hypothesis and falsely conclude that the treatment does not have an effect. Type II errors are commonly the result of a very small treatment effect. Although the treatment does have an effect, it is not large enough to show up in the research study.

Directional Tests

When a research study predicts a specific direction for the treatment effect (increase or decrease), it is possible to incorporate the directional prediction into the hypothesis test. The result is called a **directional test** or a **one-tailed test**. A directional test includes the directional prediction in the statement of the hypotheses and in the location of the critical region. For example, if the original population has a mean of $\mu = 80$ and the treatment is predicted to increase the scores, then the null hypothesis would state that after treatment:

H_0: $\mu \leq 80$ (there is no increase)

In this case, the entire critical region would be located in the right-hand tail of the distribution because large values for M would demonstrate that there is an increase and would tend to reject the null hypothesis.

Measuring Effect Size

A hypothesis test evaluates the *statistical significance* of the results from a research study. That is, the test determines whether or not it is likely that the obtained sample mean occurred without any contribution from a treatment effect. As we demonstrated in Example 8.5 in the textbook (page 261), the hypothesis test is influenced not only by the size of the treatment effect but also by the size of the sample. Thus, even a very small effect can be significant if it is observed in a very large sample. Because a significant effect does not necessarily mean a large effect, it is recommended that the hypothesis test be accompanied by a measure of the **effect size**. In this chapter, we introduce Cohen's d as a standardized measure of effect size. Much like a z-score, **Cohen's d** measures the size of the mean difference in terms of the standard deviation.

Power of a Hypothesis Test

The **power** of a hypothesis test is defined is the probability that the test will reject the null hypothesis when the treatment does have an effect. The power of a test depends on a variety of factors including the size of the treatment effect and the size of the sample.

LEARNING OBJECTIVES

1. Understand the logic of hypothesis testing.

2. Be able to state the hypotheses and find the critical region.

3. Be able to assess sample data with a z-score and make a statistical decision about the hypotheses.

4. Understand the errors that can occur in a hypothesis test: Type I and Type II errors.

5. When an experiment contains a prediction about the direction of a treatment effect, you should be able to incorporate the directional prediction into the hypothesis testing procedure and conduct a directional (one-tailed) hypothesis test.

6. Understand why a measure of effect size should accompany a hypothesis and be able to compute Cohen's d and understand what it measures.

7. Understand the concept of power for a hypothesis test and recognize the factors that influence power.

NEW TERMS AND CONCEPTS

The following terms were introduced in Chapter 8. You should be able to define or describe each term and, where appropriate, describe how each term is related to other terms in the list.

Hypothesis testing	A statistical procedure that uses data from a sample to test a hypothesis about a population.
Null hypothesis, H_0	The null hypothesis states that there is no effect, no difference, or no relationship.
Alternative hypothesis, H_1	The alternative hypothesis states that there is an effect, there is a difference, or there is a relationship.
Type I error	A Type I error is rejecting a true null hypothesis. You have concluded that a treatment does have an effect when actually it does not.
Type II error	A Type II error is failing to reject a false null hypothesis. The test fails to detect a real treatment effect.
alpha (α)	Alpha is a probability value that defines the *very unlikely* outcomes if the null hypothesis is true. Alpha also is the probability of committing a Type I error.
Level of significance	The level of significance is the alpha level, which measures the probability of a Type I error.
Critical region	The critical region consists of outcomes that are very unlikely to be obtained if the null hypothesis is true. The term *very unlikely* is defined by α.
Test statistic	A statistic that summarizes the sample data in a hypothesis test. The test statistic is used to determine whether or not the data are in the critical region.

beta (β) Beta is the probability of a Type II error.

Directional (one-tailed) test A directional test is a hypothesis test that includes a directional prediction in the statement of the hypotheses and places the critical region entirely in one tail of the distribution.

Effect size A measure of the size of the treatment effect that is separate from the statistical significance of the effect

Power The probability that the hypothesis test will reject the null hypothesis when there actually is a treatment effect.

NEW FORMULAS

$$p(\text{Type I Error}) = \alpha$$

$$p(\text{Type II Error}) = \beta$$

$$\text{Cohen's d} = \frac{\text{Mean Difference}}{\text{Standard Deviation}} = \frac{M - \mu}{\sigma}$$

STEP BY STEP

<u>Using a Sample to Test a Hypothesis about a Population Mean</u>: Although the hypothesis testing procedure is presented repeatedly in the textbook, we will demonstrate one more example here. As always, we will use the standard four-step procedure. The following generic example will be used for this demonstration.

The researcher begins with a known population, in this case a normal distribution with $\mu = 50$ and $\sigma = 10$. The researcher suspects that a particular treatment will produce a change in the scores for the individuals in the population. Because it is impossible to administer the treatment to the entire population, a sample of n = 25 individuals is selected and the treatment is given to the sample. After receiving the treatment, the average score for the sample is M = 53. Although the experiment involves only a sample, the researcher would like to use the data to make a general conclusion about how the treatment affects the entire population.

Step 1: The first step is to state the hypotheses and select an alpha level. The hypotheses always concern an unknown population. For this example, the researcher does not know what would happen if the entire population were given the treatment. Nonetheless, it is possible to state hypotheses about the effect of the treatment. Specifically, the null hypothesis says that the treatment has no effect. According to H_0, the unknown population (after treatment) is identical to the original population (before treatment). In symbols,

H_0: $\mu = 50$ (After treatment, the mean is still 50)

The alternative to the null hypothesis is that the treatment does have an effect that causes a change in the population mean. In symbols,

H_1: $\mu \neq 50$ (After treatment, the mean is different from 50)

At this time you also select the alpha-level. Traditionally, α is set at .05 or .01. If there is particular concern about a Type I error, or if a researcher desires to present overwhelming evidence for a treatment effect, a smaller alpha-level can be used (such as $\alpha = .001$).

Step 2: The next step is to locate the critical region. You should recall that the critical region is defined as the set of outcomes that are very unlikely to be obtained if the null hypothesis is true. We begin by looking at all the possible outcomes that could be obtained, then use the alpha level to determine the outcomes that are very unlikely. For this example, we look at the distribution of sample means for n = 25; that is, all the possible sample means that could be obtained if H_0 were true.

The distribution of sample means will be normal because the original population is normal. The expected value is $\mu = 50$ (if H_0 is true), and the standard error for n = 25 is

$$\sigma_M = \frac{\sigma}{\sqrt{n}} = \frac{10}{\sqrt{25}} = \frac{10}{5} = 2$$

With $\alpha = .05$, we want to identify the most unlikely 5% of this distribution. The boundaries for the extreme 5% are determined by z-scores of $z = \pm 1.96$.

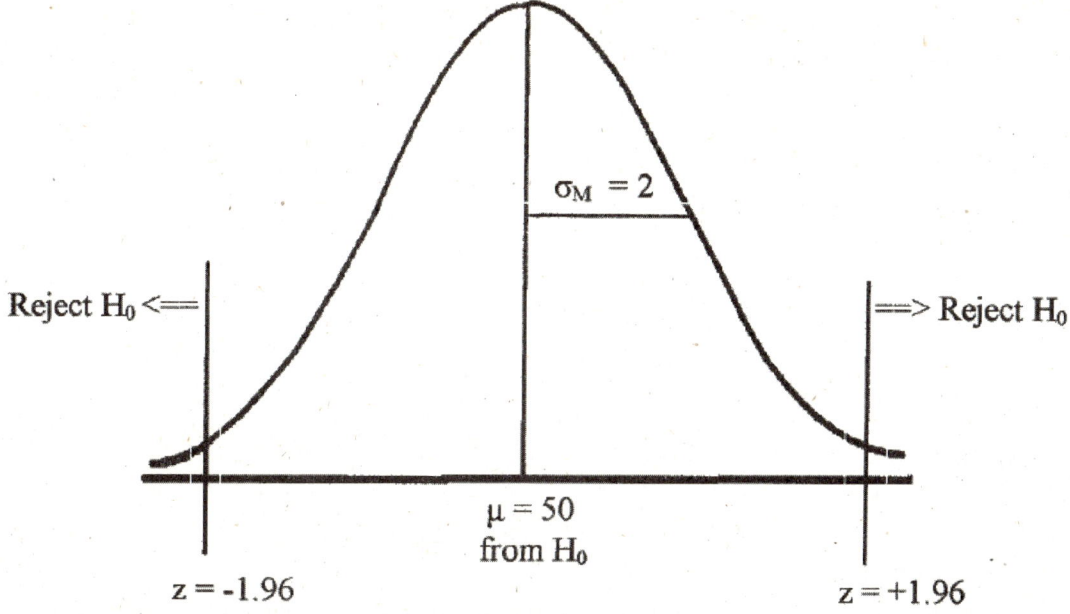

Step 3: Obtain the Sample Data and Compute the Test Statistic. For this example we obtained a sample mean of M = 53. This sample mean corresponds to a z-score of,

$$z = \frac{M - \mu}{\sigma_M} = \frac{53 - 50}{2} = \frac{3}{2} = 1.50$$

Step 4: Make Your Decision. The z-score we obtained is not in the critical region. This means that our sample mean, M = 53, is not an extreme or unusual value to be obtained from a population with $\mu = 50$. Therefore, we conclude that this sample does not provide sufficient evidence to conclude that the null hypothesis is wrong. Our statistical decision is to fail to reject H_0. The conclusion for the experiment is that the data do not indicate that the treatment has a significant effect. Note that the decision always consists of two parts:
   1) a statistical decision about the null hypothesis, and
   2) a conclusion about the outcome of the experiment

<u>Measuring Effect Size with Cohen's d</u>: Although effect size is more commonly measured after a hypothesis test in which H_0 is rejected, we will compute Cohen's d for the data from the preceding example. The original (untreated) population had $\mu = 50$. After treatment, the sample had a mean of M = 53, and the standard deviation was $\sigma = 10$. Using these values in the equation for Cohen's d, we obtain

$$\text{Cohen's d} = \frac{M - \mu}{\sigma} = \frac{53 - 50}{10} = 0.30$$

According to the criteria in Table 8.2, this is a small treatment effect.

HINTS AND CAUTIONS

1. When using samples with n > 1, we compute a z-score for the sample mean to determine if the sample data are unlikely. Be sure to use the standard error σ_M in the denominator, because the z-score is locating the sample mean within the distribution of sample means.

2. When stating the hypotheses for a directional test, remember that the predicted outcome (an increase or a decrease) is stated in the alternative hypothesis (H_1).

SELF TEST

True/False Questions

1. In general, the null hypothesis states that the treatment has no effect on the population mean.

2. In general, the null hypothesis states that the sample mean (after treatment) is equal to the original population mean (before treatment).

3. In a hypothesis test, the critical region consists of sample outcomes that have a high probability of occurring if the null hypothesis is true.

4. If the obtained sample data are in the critical region, then the correct decision is to reject the null hypothesis.

5. A Type I error occurs when a researcher concludes that a treatment has an effect but, in fact, the treatment has no effect.

6. The alpha level determines the risk of a Type I error.

7. If a sample is located in the critical region with $\alpha = .01$, then the sample would definitely be in the critical region if alpha is changed to $\alpha = .05$.

8. If a hypothesis test results in rejecting the null hypothesis, then the researcher must conclude that the treatment "does not have a significant effect."

9. In general, the larger the variance (or standard deviation), the more likely you are to reject the null hypothesis.

10. Although the size of the sample can influence the outcome of a hypothesis test, it has little or no influence on measures of effect size.

Multiple-Choice Questions

1. By selecting a smaller alpha level, a researcher is _____.
    a. attempting to make it easier to reject H_0
    b. better able to detect a treatment effect
    c. reducing the risk of a Type I error
    d. All of the above

2. If the alpha level is increased from $\alpha = .01$ to $\alpha = .05$, what happens to the size of the critical region.
    a. In increases.
    b. It decreases.
    c. The alpha level has no influence on the size of the critical region.

3. If the sample size is increased from n = 20 to n = 50, what happens to the size of the critical region.
    a. In increases.
    b. It decreases.
    c. The sample size has no influence on the size of the critical region.

4. A hypothesis test is being used to evaluate a treatment effect with $\alpha = .05$. If the sample data produce a z-score of $z = -2.24$, then what is the correct decision?
    a. Reject the null hypothesis and conclude that the treatment has no effect.
    b. Reject the null hypothesis and conclude that the treatment has an effect.
    c. Fail to reject the null hypothesis and conclude that the treatment has no effect.
    d. Fail to reject the null hypothesis and conclude that the treatment has an effect.

5. The critical boundaries for a hypothesis test are $z = +1.96$ and -1.96. If the z-score for the sample data is $z = -1.90$, then what is the correct statistical decision?
    a. Fail to reject H_1.
    b. Fail to reject H_0.
    c. Reject H_1.
    d. Reject H_0.

6. In a hypothesis test, the critical region consist of
    a. sample values that are very unlikely to occur if the null hypothesis is true
    b. sample values that are very likely to occur if the null hypothesis is true
    c. sample values that are very unlikely to occur if the null hypothesis is false
    d. sample values that prove that the null hypothesis is true

7. A Type I error means that a researcher has _____.
    a. falsely concluded that a treatment has an effect
    b. correctly concluded that a treatment has no effect
    c. falsely concluded that a treatment has no effect
    d. correctly concluded that a treatment has an effect

8. If a treatment has a very small effect, then a hypothesis test is likely to
    a. result in a Type I error.
    b. result in a Type II error.
    c. correctly reject the null hypothesis.
    d. correctly fail to reject the null hypothesis.

9. A population is known to have a mean of $\mu = 45$. If a researcher predicts that the experimental treatment will produce a *decrease* in the scores, then the null hypothesis for a one-tailed test would state _____.
    a. $\mu \geq 45$
    b. $\mu \leq 45$
    c. $M \geq 45$
    d. $M \leq 45$

10. If a hypothesis test produce a z-score in the critical region, how should the results be reported in the literature?
    a. a significant treatment effect with $p > .05$
    b. a significant treatment effect with $p < .05$
    c. no significant treatment effect with $p > .05$
    d. no significant treatment effect with $p < .05$

11. A researcher administers a treatment to a sample of n = 25 participants and uses a hypothesis test to evaluate the effect of the treatment. The hypothesis test produces a z-score of z = 2.77. Assuming that the researcher is using a two-tailed test,
    a. The researcher rejects the null hypothesis with $\alpha = .05$ but not with $\alpha = .01$.
    b. The researcher should reject the null hypothesis with either $\alpha = .05$ or $\alpha = .01$.
    c. The researcher should fail to reject H_0 with either $\alpha = .05$ or $\alpha = .01$.
    d. Cannot answer without additional information.

12. If other factors are held constant, which combination is most likely to result in rejecting the null hypothesis?
    a. a large sample size and a large population standard deviation
    b. a large sample size and a small population standard deviation
    c. a small sample size and a large population standard deviation
    d. a small sample size and a small population standard deviation

13. A treatment is administered to a sample of n = 25 selected from a population with a mean of $\mu = 80$ and a standard deviation of $\sigma = 10$. After treatment, the sample mean is M = 85. If the effect size is measured by Cohen's d, what is the value of d?
    a. d = 5.00
    b. d = 2.50
    c. d = 0.50
    d. impossible to calculate without more information

14. The power of a hypothesis test is defined as
    a. the probability that the test will reject H_0 if there is a real treatment effect.
    b. the probability that the test will fail to reject H_0 if the treatment has no effect.
    c. the probability that the test will reject H_0 if the treatment has no effect.
    d. the probability that the test will fail to reject H_0 if there is a real treatment effect.

15. Which of the following will increase the power of a hypothesis test?
    a. change α from .05 to .01
    b. change the sample size from 25 to 100
    c. change from a one-tailed test to a two-tailed test
    d. none of the other options will increase power

Other Questions

1. There is always some probability that a hypothesis test will lead to the wrong conclusion.
    a. Define a Type I error.
    b. Describe the consequences of a Type I error.
    c. What determines the probability of a Type I error.
    d. Define a Type II error.
    e. Describe the consequences of a Type II error.

2. Some researchers claim that herbal supplements such as ginseng or ginkgo biloba enhance human memory. To test this claim, a researcher selects a sample of n = 25 college students. Each student is given a ginkgo biloba supplement daily for six weeks and then all the participants are given a standardized memory test. Scores on the test are normally distributed with $\mu = 70$ and $\sigma = 15$. The sample of n = 25 students had a mean score of M = 75. Is this sample sufficient to conclude that the herb has a significant effect on memory? Use a two-tailed test with $\alpha = .05$.

3. A researcher would like to determine whether an over-the-counter cold medication has an effect on mental alertness. A sample of n = 16 participants is obtained, and each person is given a standard dose of the medication one hour before being tested in a driving simulation task. For the general population, reaction time scores on the simulation task are normally distributed with $\mu = 210$ and $\sigma = 20$. The individuals in the sample had an average score of M = 222.
    a. Can the research conclude that the medication has a significant effect on mental alertness as measured by the driving simulation task? Use a two-tailed test with $\alpha = .05$.
    b. Compute Cohen's d to measure the size of the effect.

ANSWERS TO SELF TEST

True/False Answers

1. True
2. True
3. False. The critical region consists of sample means with very low probability of occurring if the null hypothesis is true.
4. True
5. True
6. True
7. True
8. False. Rejecting the null hypothesis indicates that the treatment does have an effect.
9. False. Large variance lowers the likelihood of rejecting the null hypothesis.
10. True

Multiple-Choice Answers

1. c The alpha level is the probability of a Type I error. The smaller the alpha, the smaller the probability.
2. a Increasing α moves the critical boundaries closer to the mean.
3. c The critical region is determined by the alpha level, not the sample size.
4. b The z-score is in the critical region so reject the null hypothesis and conclude that the treatment has a significant effect.
5. b The z-score is not in the critical region so fail to reject the null hypothesis.
6. a This is a definition of the *critical region*.
7. a A Type I error means that you conclude that the treatment has an effect, when it really does not.
8. b If the treatment effect is very small, there is a good chance that the hypothesis test will fail to detect it (which is a Type II error).
9. a The null hypothesis says that there will be *no decrease*. (The population mean is still 45, or even larger.)
10. b A value in the critical region is very unlikely and significant.
11. b The z-score exceeds the 1.96 value for .05 and the 2.58 value for .01.
12. b A large sample and a small variance produce a small standard error
13. c Cohen's d is the mean difference (5) divided by the standard deviation (10).
14. a This is a definition of power.
15. b As sample size increases, so does the power of a test.

Other Answers

1. a. A Type I error is rejecting a true null hypothesis.
   b. With a Type I error, a researcher concludes that a treatment has an effect when in fact it does not. This can lead to a false report.
   c. The probability of a Type I error is the alpha level selected by the researcher.
   d. A Type II error is failing to reject a false null hypothesis.
   e. With a Type II error, a researcher concludes that the data do not provide a convincing demonstration that the treatment has any effect, when there is an effect. The researcher may choose to refine and repeat the experiment.

2. The critical region consists of values greater than $z = +1.96$ or less than $z = -1.96$. For this sample, the standard error is 3 and $z = 1.67$. Fail to reject the null hypothesis and conclude that this sample does not support the claim that the herb enhances memory.

3. a. The null hypothesis states that the medication has no effect: Stated in symbols, $H_0: \mu = 210$. The critical region consists of z-scores beyond 1.96 or −1.96. With a standard error of 5 points, the data produce a z-score of $z = 2.40$. Reject H_0 and conclude that medication has a significant effect on alertness.
   b. Cohen's $d = 12/20 = 0.60$, a medium effect according to the criteria in Table 8.2.

CHAPTER 9

INTRODUCTION TO THE t STATISTIC

CHAPTER SUMMARY

The goals of Chapter 9 are:
1. To introduce the t statistic as an alternative to the z-score hypothesis test.
2. To demonstrate the calculation of estimated standard error and the t statistic.
3. To introduce the t distributions and their relation to degrees of freedom.
4. To demonstrate the entire process of hypothesis testing with the t statistic.
5. To demonstrate the measurement of effect size for the t statistic.

The t Statistic

The t statistic introduced in Chapter 9 allows researchers to use sample data to test hypotheses about an unknown population mean. The particular advantage of the t statistic, compared to the z-score test in Chapter 8, is that the t statistic does not require any knowledge of the population standard deviation. Thus, the t statistic can be used to test hypotheses about a *completely unknown* population; that is, both μ and σ are unknown, and the only available information about the population comes from the sample. All that is required for a hypothesis test with t is a sample and a reasonable hypothesis about the population mean. There are two general situations where this type of hypothesis test is used:
1. The t statistic is used when a researcher wants to determine whether or not a treatment causes a change in a population mean. In this case you must know the value of μ for the original, untreated population. A sample is obtained from the population and the treatment is administered to the sample. If the resulting sample mean is significantly different from the original population mean, you can conclude that the treatment has a significant effect.
2. Occasionally a theory or other prediction will provide a hypothesized value for an unknown population mean. A sample is then obtained from the population and the t statistic is used to compare the actual sample mean with the hypothesized population mean. A significant difference indicates that the hypothesized value for μ should be rejected.

The Estimated Standard Error and the t Statistic

Whenever a sample is obtained from a population you expect to find some discrepancy or "error" between the sample mean and the population mean. This general phenomenon is known as **sampling error**, and was introduced in Figure 1.2 in your textbook. The goal for a hypothesis test is to evaluate the *significance* of the observed discrepancy between a sample mean and the population mean. Specifically, the hypothesis test attempts to decide between the following two alternatives:
1. Is it reasonable that the discrepancy between M and μ is simply due to sampling error and not the result of a treatment effect?

2. Is the discrepancy between M and μ more than would be expected by sampling error alone? That is, is the sample mean significantly different from the population mean?

The critical first step for the t statistic hypothesis test is to calculate exactly how much difference between M and μ is reasonable to expect. However, because the population standard deviation is unknown, it is impossible to compute the standard error of M as we did with z-scores in Chapter 8. Therefore, the t statistic requires that you use the sample data to compute an **estimated standard error of M**. This calculation defines standard error exactly as it was defined in Chapters 7 and 8, but now we must use the sample variance, s^2, in place of the unknown population variance, σ^2 (or use sample standard deviation, s, in place of the unknown population standard deviation, σ). The resulting formula for estimated standard error is

$$s_M = \sqrt{\frac{s^2}{n}} \quad \text{or} \quad s_M = \frac{s}{\sqrt{n}}$$

The t statistic (like the z-score) forms a ratio. The top of the ratio contains the obtained difference between the sample mean and the hypothesized population mean. The bottom of the ratio is the standard error which measures how much difference is expected by chance.

$$t = \frac{\text{obtained difference}}{\text{standard error}} = \frac{M - \mu}{s_M}$$

A large value for t (a large ratio) indicates that the obtained difference between the data and the hypothesis is greater than would be expected if the treatment has no effect.

The t Distributions and Degrees of Freedom

You should realize that the t statistic is very similar to the z-score used for hypothesis testing in Chapter 8. In fact, you can think of the t statistic as an "estimated z-score." The estimation comes from the fact that we are using the sample variance to estimate the unknown population variance. With a large sample, the estimation is very good and the t statistic will be very similar to a z-score. With small samples, however, the t statistic will provide a relatively poor estimate of z. The value of **degrees of freedom**, $df = n - 1$, is used to describe how well the t statistic represents a z-score. Also, the value of df will determine how well the distribution of t approximates a normal distribution. For large values of df, the **t distribution** will be nearly normal, but with small values for df, the t distribution will be flatter and more spread out than a normal distribution.

To evaluate the t statistic from a hypothesis test, you must select an α level, find the value of df for the t statistic, and consult the t distribution table. If the obtained t statistic is larger than the critical value from the table, you can reject the null hypothesis. In this case, you have demonstrated that the obtained difference between the data and the hypothesis (numerator of the ratio) is significantly larger than the difference that would be expected if there was no treatment effect (the standard error in the denominator).

Hypothesis Tests with the t Statistic

The hypothesis test with a t statistic follows the same four-step procedure that was used with z-score tests (Chapter 8):

1. State the hypotheses and select a value for α. (Note: The null hypothesis always states a specific value for μ.)
2. Locate the critical region.
   (Note: You must find the value for df and use the t distribution table.)
3. Calculate the test statistic.
4. Make a decision (Either "reject" or "fail to reject" the null hypothesis).

Measuring Effect Size with the t Statistic

Because the significance of a treatment effect is determined partially by the size of the effect and partially by the size of the sample, you cannot assume that a significant effect is also a large effect. Therefore, it is recommended that a measure of effect size be computed along with the hypothesis test.

For the t test it is possible to compute an estimate of Cohen's d just as we did for the z-score test in Chapter 8. The only change is that we now use the sample standard deviation instead of the population value (which is unknown).

$$\text{estimated Cohen's d} = \frac{\text{mean difference}}{\text{standard deviation}} = \frac{M - \mu}{s}$$

As before, Cohen's d measures the size of the treatment effect in terms of the standard deviation.

With a t test it is also possible to measure effect size by computing the **percentage of variance accounted for** by the treatment. This measure is based on the idea that the treatment causes the scores to change, which contributes to the observed variability in the data. By measuring the amount of variability that can be attributed to the treatment, we obtain a measure of the size of the treatment effect. For the t statistic hypothesis test,

$$\text{percentage of variance accounted for} = r^2 = \frac{t^2}{t^2 + df}$$

LEARNING OBJECTIVES

1. Know when you must use the t statistic rather than a z-score for hypothesis testing.

2. Understand and be able to compute the estimated standard error and the t statistic for a sample mean.

3. Understand the concept of degrees of freedom and how it relates to the t distribution.

4. Be able to perform all of the necessary computations for hypothesis tests with the t statistic.

5. Be able to compute either an estimate of Cohen's d or the value for r^2 (the percentage of variance accounted for) to measure effect size for the t test.

NEW TERMS AND CONCEPTS

The following terms were introduced in this chapter. You should be able to define or describe each term and, where appropriate, describe how each term is related to other terms in the list.

t statistic	A statistic used to summarize sample data in situations where the population standard deviation is not known. The t statistic is similar to a z-score for a sample mean, but the t statistic uses an estimate of the standard error.
Estimated standard error	An estimate of the standard error that uses the sample variance (or standard deviation) in place of the corresponding population value.
Degrees of freedom (df)	Degrees of freedom = df = n - 1, measures the number of scores that are free to vary when computing SS for sample data. The value of df also describes how well a t statistic estimates a z-score.
t distribution	The distribution of t statistics is symmetrical and centered at zero like a normal distribution. A t distribution is flatter and more spread out than the normal distribution, but approaches a normal shape as df increases.
(r^2) the percentage of variance accounted for	A measure of effect size that determines what portion of the variability in the scores can be accounted for by the treatment effect.

NEW FORMULAS

$$t = \frac{M - \mu}{s_M}$$

$$s_M = \sqrt{\frac{s^2}{n}} \quad \text{or} \quad s_M = \frac{s}{\sqrt{n}}$$

$$\text{percentage of variance accounted for} = r^2 = \frac{t^2}{t^2 + df}$$

STEP BY STEP

<u>Hypothesis Testing with the t Statistic</u>: The t statistic presented in this chapter is used to test a hypothesis about an unknown population mean using the data from a single sample. Calculation of the t statistic requires the sample mean M and some measure of the sample variability, usually the sample variance, s^2. A hypothesis test with the t statistic uses the same four-step procedure that we use for all hypothesis tests. However, with a t statistic, you must compute the variance (or standard deviation) for the sample of scores and you must remember to use the t distribution table to locate the critical values for the test. We will use the following example to demonstrate the t statistic hypothesis test.

A psychologist has prepared an "Optimism Test" that is administered yearly to graduating college seniors. The test measures how each graduating class feels about its future -- the higher the score, the more optimistic the class. Last year's class had a mean score of $\mu = 56$. A sample of $n = 25$ seniors from this year's class produced an average score of $M = 59$ with $SS = 2400$. On the basis of this sample can the psychologist conclude that this year's class has a different level of optimism than last year's class? Test at the .05 level of significance.

Note that this test will use a t statistic because the population standard deviation is not known.

Step 1: State Hypotheses and select an alpha level.
The statements of the null hypothesis and the alternative hypothesis are the same for the t statistic test as they were for the z-score test.

H_0: $\mu = 56$ (no change)
H_1: $\mu \neq 56$ (this year's mean is different)

For this example we are using $\alpha = .05$

Step 2: Locate the critical region. With a sample of n = 25, the t statistic will have df = 24. For a two-tailed test with α = .05 and df = 24, the critical t values are t = ∓2.064.

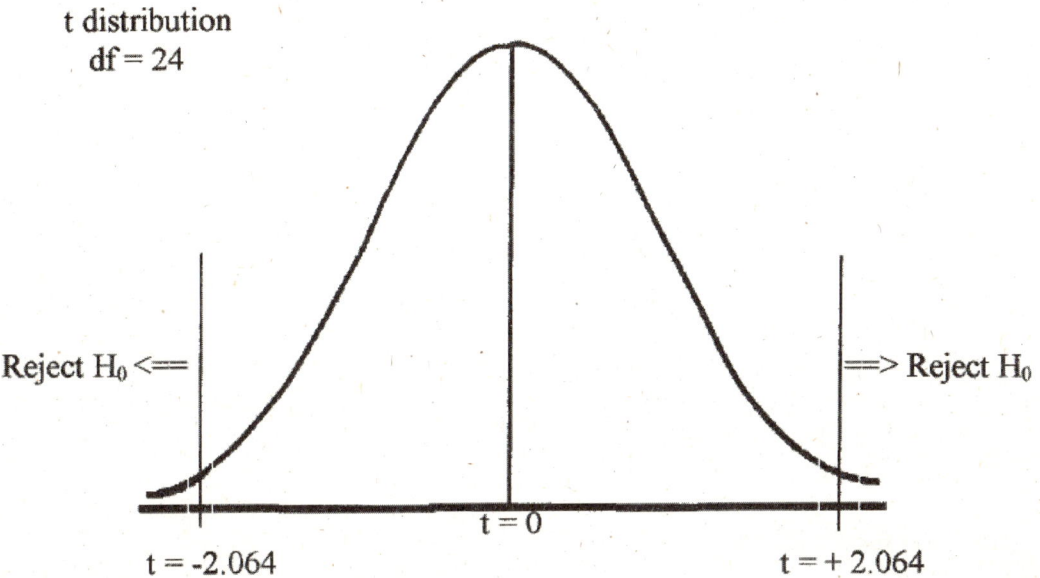

Step 3: Obtain the data and compute the test statistic: For this sample we have M = 59, SS = 2400, and n = 25. To compute the t statistic for these data, it is best to start by calculating the sample variance.

$$s^2 = \frac{SS}{n-1} = \frac{2400}{24} = 100$$

Next, use the sample variance to compute the estimated standard error. Remember, standard error provides a measure of the standard distance between a sample mean M and its population mean μ.

$$s_M = \sqrt{\frac{s^2}{n}} = \sqrt{\frac{100}{25}} = \sqrt{4} = 2$$

Finally, compute the t statistic using the hypothesized value of μ from H_0.

$$t = \frac{M - \mu}{s_M} = \frac{59 - 56}{2} = \frac{3}{2} = 1.50$$

INTRODUCTION TO THE t STATISTIC

Step 4: Make decision. The t statistic we obtained is not in the critical region. Because the sample mean is reasonably close to the hypothesized population mean, we fail to reject H_0. These data do not provide sufficient evidence to conclude that this year's graduating class has a level of optimism that is different from last year's class.

<u>Measuring Effect Size for the t Statistic</u>: Effect size can be measured by estimating Cohen's d or computing r^2. For the data from the hypothesis testing example,

$$\text{Cohen's d} = \frac{M - \mu}{s} = \frac{59 - 56}{10} = \frac{3}{10} = 0.30 \quad \text{(a small to medium effect)}$$

$$r^2 = \frac{t^2}{t^2 + df} = \frac{(1.50)^2}{(1.50)^2 + 24} = \frac{2.25}{26.25} = 0.086 \quad \text{(or 8.6\%)}$$

HINTS AND CAUTIONS

1. Students often confuse the formulas for sample variance (or standard deviation) and estimated standard error. The specific confusion is deciding when to divide by n and when to divide by n - 1.

   Sample variance and standard deviation are *descriptive statistics* that were introduced in Chapter 4. You should recall that the sample mean describes the center of the distribution, and the sample standard deviation (or variance) describes how the scores are distributed around the mean. To compute the sample variance or the sample standard deviation, the correct denominator is df = n - 1.

   The estimated standard error (s_M), on the other hand, is primarily an *inferential statistic* that measures how accurately the sample mean represents its population mean. The degree of accuracy is largely determined by the size of the sample: The bigger the sample, the smaller the error. Thus, the magnitude of the standard error is determined by n. To compute the estimated standard error, the correct denominator is n.

2. When locating the critical region for a t test, be sure to consult the t distribution table, not the unit normal table.

SELF TEST

                    True/False Questions

1. The larger the value for df, the more a t distribution resembles a normal distribution.

2. In a t statistic, the estimated standard error provides a measure of how much difference is reasonable to expect between a sample mean and the population mean.

3. The size of the estimated standard error, s_M, is partially determined by the size of the sample variance.

4. As the value of df increases, the t distribution tends to become flatter and more spread out.

5. Except for the population mean, μ, all the numbers in the t statistic formula come from the sample data.

6. In a hypothesis test using the t statistic, an increase in the size of the sample variance produces an increase in the likelihood that the null hypothesis will be rejected.

7. A hypothesis test produces t = 2.062 for a sample of n = 25 scores. For a two-tailed test with α = .05, the correct decision is to reject the null hypothesis.

8. A researcher reports a significant treatment effect with t(15) = 2.56, p < .05. The study used a sample of n = 15 participants.

9. If other factors are held constant, increasing the number of scores in the sample will increase the likelihood of rejecting the null hypothesis.

10. If other factors are held constant, the value of Cohen's d increases as the sample variance increases.

Multiple-Choice Questions

1. Several samples, each with n = 9 scores, are selected from a population. If the t statistic and the z-score are computed for each sample mean, which of the following is probably true?
    a. The average t statistic will be larger than the average z-score.
    b. The average t statistic will be smaller than the average z-score.
    c. The set of t statistics will be more variable that the set of z-scores.
    d. The set of t statistics will be less variable that the set of z-scores.

2. To calculate a t statistic, what information is needed from the sample?
    a. the value for n
    b. the value for M
    c. the value for s or s^2
    d. All of the other options are needed to compute t.

3. A sample of n = 4 scores has a mean of M = 35 and SS = 48. What are the values for the sample standard deviation and the estimated standard error for the sample mean?
    a. s = 16 and s_M = 4
    b. s = 16 and s_M = 2
    c. s = 4 and s_M = 1
    d. s = 4 and s_M = 2

INTRODUCTION TO THE t STATISTIC

4. What is the estimated standard error for a sample of n = 4 scores with a variance of $s^2 = 36$?
    a. the square root of (36/4)
    b. the square root of (36/3)
    c. 36/4
    d. 36/3

5. The estimated standard error, s_M, provides a measure of the average or standard distance between _____.
    a. a score and the population mean (X and μ)
    b. a sample mean and the population mean (M and μ)
    c. a score and the sample mean (X and M)
    d. None of the other options is correct.

6. If a sample of n = 25 scores produces a t statistic of t = −2.36, which of the following decisions is justified?
    a. Reject H_0 with α = .05 but fail to reject with α = .01.
    b. Reject H_0 with α = .05 or with α = .01.
    c. Fail to reject H_0 with α = .05 and fail to reject with α = .01.
    d. Fail to reject H_0 with α = .05 but reject H_0 with α = .01.

7. A sample of n = 9 scores has a mean of M = 46 and a variance of s^2 = 36. What is the estimated standard error for this sample?
    a. 12
    b. 6
    c. 4
    d. 2

8. A sample of n = 25 scores has a mean of M = 40 and a variance of s^2 = 100. If this sample is being used to test a null hypothesis stating that μ = 43, then what is the t statistic for the sample?
    a. t = −3/20 = −0.15
    b. t = −0.30
    c. t = −3/4 = −0.75
    d. t = −3/2 = −1.50

9. A researcher reports a significant treatment effect with t(24) = 3.04. How many scores were in the sample?
    a. 23
    b. 24
    c. 25
    d. Cannot determine without additional information.

10. A sample of n = 4 individuals is obtained from a population with μ = 80. Which set of sample statistics would produce the most extreme value for t?
    a. $M = 84$ and $s^2 = 8$
    b. $M = 84$ and $s^2 = 32$
    c. $M = 88$ and $s^2 = 8$
    d. $M = 88$ and $s^2 = 32$

11. A sample of n = 25 scores is obtained from a population with μ = 80. Which of the following sets of sample statistics is most likely to reject the null hypothesis?
    a. $M = 85$ with $s^2 = 10$
    b. $M = 90$ with $s^2 = 10$
    c. $M = 85$ with $s^2 = 100$
    d. $M = 90$ with $s^2 = 100$

12. If other factors are held constant, which of the following is a consequence of increasing the sample size?
    a. An increased standard error and an increased likelihood of rejecting H_0.
    b. An increased standard error and a decreased likelihood of rejecting H_0.
    c. A decreased standard error and an increased likelihood of rejecting H_0.
    d. A decreased standard error and a decreased likelihood of rejecting H_0.

13. Which combination of factors has the greatest likelihood of rejecting the null hypothesis?
    a. A large sample size and a large sample variance.
    b. A large sample size and a small sample variance.
    c. A small sample size and a large sample variance.
    d. A small sample size and a small sample variance.

14. A sample of n = 16 individuals is selected from a population with μ = 80 and a treatment is administered to the sample. After treatment, the sample mean is M = 84 and the sample variance is $s^2 = 100$. If Cohen's d is used to measure effect size for this study, what value will be obtained for d?
    a. 4/2.5 = 1.60
    b. 0.40
    c. 0.04
    d. cannot be determined without additional information

15. A hypothesis test produces t = 2.00 for a sample of n = 16 scores. What is the value of r^2?
    a. 2/18
    b. 2/17
    c. 4/20
    d. 4/19

Other Questions

1. A research study reports t(14) = 2.17.
    a. How many individuals participated in the study? (How big was the sample?)
    b. Is this t statistic sufficient to reject the null hypothesis using a two-tailed test with $\alpha = .05$?

2. A researcher would like to evaluate the effect of a new reading program for second-grade students. For the past five years, a standardized test given at the end of second grade has produced a mean score of $\mu = 45$. A sample of n = 25 students is placed in the new program and, at the end of the school year they obtain a mean test score of M = 51 with SS = 9600. Based on these data can the researcher conclude that the new program has a significant effect on reading scores. Use a two-tailed test with $\alpha = .05$.
    a. Using symbols, state the hypotheses for this test.
    b. Locate the critical region for $\alpha = .05$.
    c. Calculate the t statistic for this sample.
    d. What decision should the researcher make?
    e. Compute Cohen's d and r^2 to measure the size of the effect.

3. A researcher obtains the following sample from a population with an unknown mean and unknown standard deviation.
    Sample: 5, 0, 4, 7, 3, 7, 8, 2, 9
    a. Compute the mean, variance, and standard deviation for the sample.
    b. Use the sample variance to compute the estimated standard error, s_M, for the sample mean.
    c. Use the sample to test the null hypothesis that the population mean is equal to $\mu = 8$. Use a two-tailed test with $\alpha = .05$. (Assume the population distribution is normal.)

ANSWERS TO SELF TEST

True/False Answers

1. True
2. True
3. True
4. True
5. True
6. False. The larger the sample variance, the less likely it is that you will reject the null hypothesis.
7. True
8. False. With df = 15, the sample size is n = 16.
9. True
10. False. Sample size has no influence on Cohen's d.

Multiple-Choice Answers

1. c Only the numerator varies with z but both the numerator and denominator vary with t.
2. d The t statistic requires the sample variance as well as M and n.
3. d The sample variance is 16.
4. a Estimated standard error equals the square root of (s^2/n).
5. b The estimated standard error measures the standard distance between M and μ.
6. a The t value is in the critical region for .05 but not for .01.
7. d Estimated standard error equals the square root of (s^2/n).
8. d The estimated standard error is 2, and $t = -3/2$.
9. c With df = 24, the sample size is n = 25.
10. c The most extreme sample has the biggest mean difference and the smallest variance.
11. b The most extreme sample has the biggest mean difference and the smallest variance.
12. c As sample size increases, the error decreases and t increases.
13. b A large sample and small variance produce the smallest standard error.
14. b The sample standard deviation is 10, and Cohen's d = 4/10.
15. d $r^2 = t^2$ divided by $(t^2 + df)$

Other Answers

1. a. With df = 14, the sample size is n = 15,
   b. With df = 14, the critical values are t = ∀2.145. Reject H_0.

2. a. H_0: μ = 45 and H_1: μ ≠ 45
   b. With df = 24, the critical values are t = ∀2.064.
   c. The sample variance is $s^2 = 400$, the standard error is 4, and t = 1.50.
   d. Fail to reject H_0. The data indicate that the new program does not have a significant effect on reading scores.
   e. Cohen's d = 6/√400 = 0.30. The percentage of variance accounted for is r^2 = 2.25/(2.25 + 24) = 0.086 (or 8.6%).

3. a. M = 5, $s^2 = 9$, and s = 3
   b. The estimated standard error is 1 point.
   c. If μ = 8 (from H_0), then t = 3.00. This is beyond the critical value of 2.306 so the decision is to reject the null hypothesis.

CHAPTER 10

THE t TEST FOR TWO INDEPENDENT SAMPLES

CHAPTER SUMMARY

The goals of Chapter 10 are:
1. To introduce the basic characteristics of an independent-measures (or between-subjects) research design.
2. To demonstrate the process of hypothesis testing with the independent-measures t statistic.
3. To introduce the assumption of homogeneity of variance.
4. To demonstrate the measurement of effect size for the independent-measures t.

Independent-Measures Designs

The independent-measures hypothesis test allows researchers to evaluate the mean difference between two populations using the data from two separate samples. The identifying characteristic of the **independent-measures** or **between-subjects** design is the existence of two separate or independent samples. Thus, an independent-measures design can be used to test for mean differences between two distinct populations (such as men versus women) or between two different treatment conditions (such as drug versus no-drug). The independent-measures design is used in situations where a researcher has no prior knowledge about either of the two populations (or treatments) being compared. In particular, the population means and standard deviations are all unknown. Because the population variances are not known, these values must be estimated from the sample data.

Hypothesis Testing with the Independent-Measures t Statistic

As with all hypothesis tests, the general purpose of the independent-measures t test is to determine whether the sample mean difference obtained in a research study indicates a real mean difference between the two populations (or treatments) or whether the obtained difference is simply the result of sampling error. Remember, if two samples are taken from the same population and are given exactly the same treatment, there still will be some difference between the sample means. This difference is called sampling error (see Figure 1.2 in your textbook). The hypothesis test provides a standardized, formal procedure for determining whether the mean difference obtained in a research study is significantly greater than can be explained by sampling error.

To prepare the data for analysis, the first step is to compute the sample mean and SS (or s, or s^2) for each of the two samples. The hypothesis test follows the same four-step procedure outlined in Chapters 8 and 9.
1. State the hypotheses and select an α level. For the independent-measures test, H_0 states that there is no difference between the two population means.

2. Locate the critical region. The critical values for the t statistic are obtained using degrees of freedom that are determined by adding together the df value for the first sample and the df value for the second sample.

3. Compute the test statistic. The t statistic for the independent-measures design has the same structure as the single sample t introduced in Chapter 9. However, in the independent-measures situation, all components of the t formula are doubled: there are two sample means, two population means, and two sources of error contributing to the standard error in the denominator.

4. Make a decision. If the t statistic ratio indicates that the obtained difference between sample means (numerator) is substantially greater than the difference expected by chance (denominator), we reject H_0 and conclude that there is a real mean difference between the two populations or treatments.

The Homogeneity of Variance Assumption

Although most hypothesis tests are built on a set of underlying assumptions, the tests usually work reasonably well even if the assumptions are violated. The one notable exception is the assumption of **homogeneity of variance** for the independent-measures t test. The assumption requires that the two populations from which the samples are obtained have equal variances. This assumption is necessary in order to justify pooling the two sample variances and using the pooled variance in the calculation of the t statistic. If the assumption is violated, then the t statistic contains two questionable values: (1) the value for the population mean difference which comes from the null hypothesis, and (2) the value for the pooled variance. The problem is that you cannot determine which of these two values is responsible for a t statistic that falls in the critical region. In particular, you cannot be certain that rejecting the null hypothesis is correct when you obtain an extreme value for t. If the two sample variances appear to be substantially different, you should use Hartley's F-max test to determine whether or not the homogeneity assumption is satisfied. If homogeneity of variance is violated, Box 10.3 presents an alternative procedure for computing the t statistic that does not involve pooling the two sample variances.

Measuring Effect Size for the Independent-Measures t

Effect size for the independent-measures t is measured in the same way that we measured effect size for the single-sample t in Chapter 9. Specifically, you can compute an estimate of Cohen's d or you can compute r^2 to obtain a measure of the percentage of variance accounted for by the treatment effect.

LEARNING OBJECTIVES

1. You should be able to describe and to recognize the experimental situations for which an independent-measures t statistic is appropriate.

2. You should be able to use the independent-measures t statistic to test hypotheses about the mean difference between two populations (or between two treatment conditions).

3. You should be able to list the assumptions that must be satisfied before an independent-measures t statistic can be computed or interpreted.

4. You should be able to use the sample data to compute r^2 or an estimate of Cohen's d to measure effect size for the independent-measures t.

NEW TERMS AND CONCEPTS

The following terms were introduced in Chapter 10. You should be able to define or describe each term and, where appropriate, describe how each term is related to others in the list.

Independent-measures design	A research design that uses a separate sample for each treatment condition or each population being compared.
Between-subjects design	An alternative term for an independent-measures design.
Repeated-measures design	A research design that uses the same group of subjects in all of the treatment conditions that are being compared.
Within-subjects design	An alternative term for a repeated-measures design.
Pooled variance	A single measure of sample variance that is obtained by averaging two sample variances. It is a weighted mean of the two variances.
Homogeneity of variance	An assumption that the two populations from which the samples were obtained have equal variances.

NEW FORMULAS

$$t = \frac{(M_1 - M_2) - (\mu_1 - \mu_2)}{s_{M_1 - M_2}} \qquad \begin{aligned} df &= (n_1 - 1) + (n_2 - 1) \\ &= n_1 + n_2 - 2 \end{aligned}$$

$$S_{M_1-M_2} = \sqrt{\frac{s_P^2}{n_1} + \frac{s_P^2}{n_2}}$$

$$s_P^2 = \frac{SS_1 + SS_2}{df_1 + df_2}$$

$$F_{max} = \frac{s^2 \text{(largest)}}{s^2 \text{(smallest)}}$$

$$\text{estimated Cohen's d} = \frac{M_1 - M_2}{\sqrt{s_P^2}}$$

STEP BY STEP

<u>Hypothesis Tests with the Independent-Measures t Statistic.</u> The independent-measures t statistic is used in situations where a researcher wants to test a hypothesis about the difference between two population means using the data from two separate (independent) samples. The test requires both sample means (M_1 and M_2), and some measure of the variability for each sample (usually SS). The following example will be used to demonstrate the independent-measures t hypothesis test.

A researcher wants to assess the damage to memory that is caused by excessive alcohol consumption. A sample of n = 10 adults with chronic alcoholism is obtained from a hospital treatment ward, and a control group of n = 10 non-drinkers is obtained from the hospital maintenance staff. Each person is given a brief memory test and the researcher records the memory score for each subject. The data are summarized as follows:

                Alcoholism   Control
                  M = 43   M = 57
                 SS = 400   SS = 410

Step 1: State the hypotheses and select an alpha level. As always, the null hypothesis states that there is no effect.

H_0: $(\mu_1 - \mu_2) = 0$ (no difference)

The alternative hypothesis says that there is a difference between the two population means.

H_1: $(\mu_1 - \mu_2) \neq 0$

We will use $\alpha = .05$.

Step 2: Locate the critical region. With n = 10 in each sample, the t statistic will have degrees of freedom equal to, $df = (n_1 - 1) + (n_2 - 1) = 9 + 9 = 18$

Sketch the entire distribution of t statistics with df = 18 and locate the extreme 5%. The critical values are t = ∀2.101.

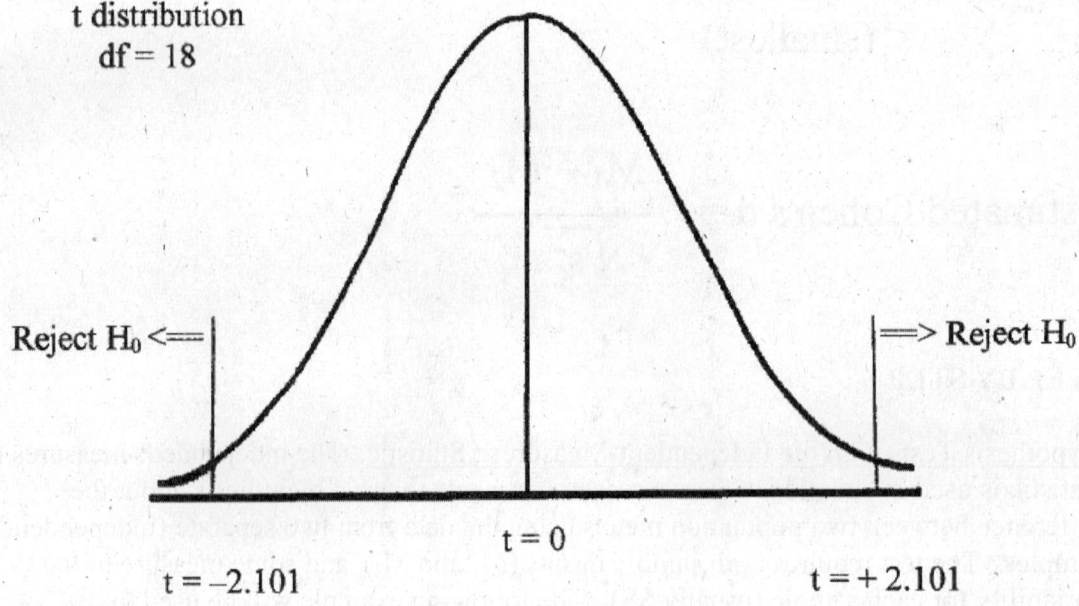

t distribution
df = 18

Reject H_0 <===   ===> Reject H_0

t = –2.101   t = 0   t = + 2.101

Step 3: Compute the t statistic. It is easiest to begin by computing the pooled variance for the two samples.

$$s_P^2 = \frac{SS_1 + SS_2}{df_1 + df_2} = \frac{400 + 410}{9 + 9} = \frac{810}{18} = 45$$

Next, calculate the standard error for the t statistic.

$$S_{M_1-M_2} = \sqrt{\frac{s_p^2}{n_1} + \frac{s_p^2}{n_2}} = \sqrt{\frac{45}{10} + \frac{45}{10}}$$

$$= \sqrt{4.5 + 4.5}$$

$$= \sqrt{9} = 3$$

Finally, use the two sample means and the standard error to calculate the t statistic.

$$t = \frac{(M_1 - M_2) - (\mu_1 - \mu_2)}{S_{M_1-M_2}} = \frac{(43 - 57) - 0}{3} = \frac{-14}{3} = -4.67$$

Step 4: Make decision. The t statistic for these data is in the critical region. This is a very unlikely outcome ($p < .05$) if H_0 is true, therefore, we reject H_0. The researcher concludes that there is a significant difference between the mean memory score for chronic alcoholics and the mean score for non-drinkers.

Measuring Effect Size for the Independent-Measures t

Effect size for the independent-measures t can be measured by estimating Cohen's d or computing r^2 (the percentage of variance accounted for). Using the data from the preceding example, these two measures are computed as follows:

$$\text{estimated Cohen's d} = \frac{M_1 - M_2}{\sqrt{s_p^2}} = \frac{43 - 57}{\sqrt{45}} = \frac{14}{6.71} = 2.09$$

$$r^2 = \frac{t^2}{t^2 + df} = \frac{(-4.67)^2}{(-4.67)^2 + 18} = \frac{21.81}{39.81} = 0.548 \quad (\text{or } 54.8\%)$$

THE t TEST FOR TWO INDEPENDENT SAMPLES

HINTS AND CAUTIONS

1. One of the most common errors in computing the independent-measures t statistic occurs when students confuse the formulas for pooled variance and standard error. To compute the pooled variance, you combine the two samples into a single estimated variance. The formula for pooled variance uses a single fraction with the SS values in the numerator and the df values in the denominator:

$$s_P^2 = \frac{SS_1 + SS_2}{df_1 + df_2}$$

To compute the standard error, you add the separate errors for the two samples. In the formula for standard error these two separate sources of error appear as two separate fractions:

$$s_{M_1 - M_2} = \sqrt{\frac{s_P^2}{n_1} + \frac{s_P^2}{n_2}}$$

SELF TEST

True/False Questions

1. If the two samples are the same size, the independent measures t statistic will have an even number for degrees of freedom.

2. Two separate samples, each with n = 10 scores, will produce an independent-measures t statistic with df = 19.

3. If the two samples are the same size, then the pooled variance will equal the average of the two sample variances.

4. One sample has n = 7 scores with SS = 40, and a second sample has n = 5 scores with SS = 80. The pooled variance for these two samples is 120/10 = 12.

5. Two samples, each with n = 4 scores, have a pooled variance of 32. The estimated standard error for the sample mean difference is √8.

6. The estimated standard error for the independent-measures t statistic provides a measure of how much difference should exist, on average, between two sample means for samples selected from the same population.

7. If other factors are held constant, the larger the values for the two sample variances, the greater the likelihood that the independent-measures t test will find a significant difference.

8. If other factors are held constant, the larger the larger the two sample sizes are, the greater the likelihood that the independent-measures t test will find a significant difference.

9. The *homogeneity of variance* assumption states that the two population variances are equal.

10. If two separate samples have $M_1 = 10$ and $M_2 = 18$ with a pooled variance of 16, then Cohen's d = 0.50.

Multiple-Choice Questions

1. Which of the following research situations would be most likely to use an independent-measures design?
    a. Examine the development of vocabulary as a group of children mature from age 2 to age 3.
    b. Examine the long-term effectiveness of a stop-smoking treatment by interviewing subjects 2 months and 6 months after the treatment ends.
    c. Compare the mathematics skills for 9th grade boys versus 9th grade girls.
    d. Compare the blood-pressure readings before medication and after medication for a group a patients with high blood pressure.

2. A researcher reports an independent-measures t statistic with df = 30. If the two samples are the same size ($n_1 = n_2$), then how many individuals are in each sample?
    a. n = 15
    b. n = 16
    c. n = 30
    d. n = 31

3. One sample of n = 10 scores has a variance of $s^2 = 10$ and a second sample of n = 10 scores has $s^2 = 20$. If the pooled variance is computed for these two samples, then the value obtained will be _____.
    a. closer to 10 than to 20
    b. closer to 20 than to 10
    c. exactly half way between 10 and 20
    d. cannot be determined without more information

4. One sample has a variance of $s^2 = 8$ and a second sample has a variance of $s^2 = 10$. If the two samples are exactly the same size ($n_1 = n_2$), then what is the pooled variance?
    a. 9
    b. 18
    c. cannot be determined without more information

5. One sample has n = 15 scores with SS = 54 and a second sample has n = 7 scores with SS = 26. What is the pooled variance for the two samples?
   a. 40/22
   b. 80/22
   c. 40/20
   d. 80/20

6. One sample has n = 8 scores with SS = 180 and a second sample has n = 4 scores with SS = 60. What is the estimated standard error for the sample mean difference?
   a. 240
   b. 24
   c. $\sqrt{24}$
   d. 3

7. What is the pooled variance for the following two samples?
   a. 5
   b. $\sqrt{5}$
   c. 12
   d. 9.6

   Sample 1: n = 6 and SS = 56
   Sample 2: n = 4 and SS = 40

8. Two samples, each with n = 5 scores, have a pooled variance of 40. What is the estimated standard error for the sample mean difference?
   a. 4
   b. 8
   c. 10
   d. $\sqrt{20}$

9. As the sample variance increases, the value of the t statistic _____.
   a. increases (moves away from zero)
   b. decreases (moves toward zero)
   c. approximates the value of $M_1 - M_2$
   d. the value of t is not influenced by sample variance

10. Two samples, each with n = 8, produce an independent-measures t statistic of t = –2.15. Which of the following decisions is justified?
    a. Reject H_0 with α = .05 but fail to reject with α = .01.
    b. Reject H_0 with α = .05 and reject H_0 with α = .01.
    c. Fail to reject H_0 with α = .05 and fail to reject with α = .01.
    d. Fail to reject H_0 with α = .05 but reject H_0 with α = .01.

11. Which of the following sets of data would produce the largest value for an independent-measures t statistic?
    a. The two sample means are 10 and 12 with sample variances of 20 and 25.
    b. The two sample means are 10 and 12 with variances of 120 and 125.
    c. The two sample means are 10 and 20 with variances of 20 and 25.
    d. The two sample means are 10 and 20 with variances of 120 and 125.

12. If other factors are held constant, which of the following sets of data would produce the largest value for an independent-measures t statistic?
    a. The two samples both have n = 15 with sample variances of 20 and 25.
    b. The two samples both have n = 15 with variances of 120 and 125.
    c. The two samples both have n = 30 with sample variances of 20 and 25.
    d. The two samples both have n = 30 with variances of 120 and 125.

13. Which set of characteristics is most likely to produce a significant difference for an independent-measures t test?
    a. large sample sizes and large sample variances
    b. large sample sizes and small sample variances
    c. small sample sizes and large sample variances
    d. small sample sizes and small sample variances

14. Two samples, each with n = 9 scores, produce an independent-measures t statistic of t = 2.00. If the effect size is measured using r^2, what is the value of r^2?
    a. 4/16
    b. 4/20
    c. 2/16
    d. 2/18

15. One sample has M = 18 and a second sample has M = 14. If the pooled variance for the two samples is 16, then what is the value of Cohen's d?
    a. 0.25
    b. 0.50
    c. 1.00
    d. cannot be determined with the information given

Other Questions

1. A researcher would like to demonstrate how different schedules of reinforcement can influence behavior. Two separate groups of rats are trained to press a bar in order to receive a food pellet. One group is trained using a fixed ratio schedule where they receive one pellet for every 10 presses of the bar. The second group is trained using a fixed interval schedule where they receive one pellet for the first bar press that occurs within a 30 second interval. Note that the second group must wait 30 seconds before another pellet is possible no matter how many times the bar is pressed. After 4 days of training,
the researcher records the response rate (number of presses per minute) for each rat. The results are summarized as follows:

Fixed Ratio	Fixed Interval
n = 4	n = 8
M = 30	M = 18
SS = 90	SS = 150

a. Do these data indicate that there is a significant difference in responding for these two reinforcement schedules? Test at the .05 level of significance.
b. Compute Cohen's d and r^2 to measure the size of the effect.

2. A psychologist is examining the influence of an older sibling in the development of social skills. A sample of 24 three-year-old children is obtained. Half of these children had no siblings and the others had at least one older sibling who is within 5 years of the child's age. The psychologist records a social skills score for each child and obtained the following data:

No Sibling	Older Sibling
n = 12	n = 12
M = 17	M = 24
SS = 580	SS = 608

Do these data indicate that having an older sibling has a significant effect on the development of social skills? Use a two-tailed test at the .05 level.

3. The following data are from two separate samples. Does it appear that these two samples came from the same population or from two different populations?
a. Use an F-max test to determine whether there is evidence for a significant difference between the two population variances. Use the .05 level of significance.
b. Use an independent-measures t test to determine whether there is evidence for a significant difference between the two population means. Again, use $\alpha = .05$.

Sample 1	Sample 2
n = 10	n = 10
M = 32	M = 18
SS = 890	SS = 550

ANSWERS TO SELF TEST

True/False Answers

1. True
2. False. df = 18
3. True
4. True
5. False. The estimated standard error is $\sqrt{16} = 4$
6. True
7. False. Increases in variance produce a smaller value for t (nearer to zero).
8. True
9. True
10. False. The mean difference is divided by the square root of the pooled variance.

Multiple-Choice Answers

1. c Comparing boys and girls would require two separate samples.
2. a The t statistic has df = $(n_1 - 1) + (n_2 - 1)$.
3. c If the two samples are the same size, the pooled variance is the average of the two sample variances.
4. a If the two samples are the same size, the pooled variance is the average of the two sample variances.
5. d Pooled variance equals $(SS_1 + SS_2)/(df_1 + df_2)$.
6. d The pooled variance is 24 and the standard error is $\sqrt{9} = 3$.
7. c Pooled variance equals $(SS_1 + SS_2)/(df_1 + df_2)$
8. a The estimated standard error is $\sqrt{16} = 4$.
9. b Increases in variance produce increases in the standard error and a smaller value for t (nearer to zero).
10. a The t is in the critical region for .05 (2.145) but not for 01 (2.977)
11. c The greatest difference between samples occurs with a big mean difference and small variances.
12. c The greatest difference between samples occurs with large sample sizes and small variances.
13. b The greatest difference between samples occurs with large sample sizes and small variances.
14. b $r^2 = t^2/(t^2 + df)$
15. c The standard deviation is estimated with the square root of the pooled variance. Cohen's d is the mean differences divided by the standard deviation.

Other Answers

1. a. With df = 10, the critical t values are t = ∀2.228. These data have a pooled variance of 24 and produce a t statistic of t = 4.00. Reject H_0. The data provide sufficient evidence to conclude that there is a significant difference between the two schedules.
   b. Cohen's d = $12/\sqrt{24}$ = 2.45 and r = 16/(16 + 10) = 0.615 (or 61.5%)

2. The hypotheses are,
   H_0: $(\mu_1 - \mu_2) = 0$ (sibling has no effect)
   H_1: $(\mu_1 - \mu_2) \neq 0$ (sibling does have an effect)
   With df = 22, the critical region consists of t values beyond t = ±2.074. For these data, the pooled variance is 54, and t = 2.33. Reject H_0 and conclude that the children with siblings have significantly different social skills scores than those without siblings.

3. a. For these data, F-max = 1.62. The critical value for α = .05 is 4.03. Fail to reject H_0. There is insufficient evidence to conclude that the two population variances are different.
   b. For these data the pooled variance is 80 and the t statistic is t = 14/4 = 3.50. Reject H_0 and conclude that the two population means are different.

CHAPTER 11

THE t TEST FOR TWO RELATED SAMPLES

CHAPTER SUMMARY

The goals of Chapter 11 are:
1. To introduce the basic characteristics of a repeated-measures (or within-subjects) research design.
2. To demonstrate the process of hypothesis testing with the repeated-measures t statistic.
3. To demonstrate the measurement of effect size for the repeated-measures t
4. To discuss the advantages and the disadvantages of a repeated-measures design compared to an independent-measures design.

Repeated-Measures Designs
 The related-samples hypothesis test allows researchers to evaluate the mean difference between two treatment conditions using the data from a single sample. In a **repeated-measures design**, a single group of individuals is obtained and each individual is measured in both of the treatment conditions being compared. Thus, the data consist of two scores for each individual.

Hypothesis Tests with the Repeated-Measures t
 The repeated-measures t statistic allows researchers to test a hypothesis about the population mean difference between two treatment conditions using sample data from a repeated-measures research study. In this situation it is possible to compute a **difference score** for each individual:

$$\text{difference score} = D = X_2 - X_1$$

Where X_1 is the person's score in the first treatment and X_2 is the score in the second treatment.

Note: The related-samples t test can also be used for a similar design, called a matched-subjects design, in which each individual in one treatment is matched one-to-one with a corresponding individual in the second treatment. The matching is accomplished by selecting pairs of subjects so that the two subjects in each pair have identical (or nearly identical) scores on the variable that is being used for matching. Thus, the data consist of pairs of scores with each pair corresponding to a matched set of two "identical" subjects. For a matched-subjects design, a difference score is computed for each matched pair of individuals. However, because the matching process can never be perfect, matched-subjects designs are relatively rare. As a result, repeated-measures designs (using the same individuals in both treatments) make up the vast majority of related-samples studies and we will focus on them.

The sample of difference scores is used to test hypotheses about the population of difference scores. The null hypothesis states that the population of difference scores has a mean of zero,

$$H_0: \mu_D = 0$$

In words, the null hypothesis says that there is no consistent or systematic difference between the two treatment conditions. Note that the null hypothesis does not say that each individual will have a difference score equal to zero. Some individuals will show a positive change from one treatment to the other, and some will show a negative change. On average, however, the entire population will show a mean difference of zero. Thus, according to the null hypothesis, the sample mean difference should be near to zero. Remember, the concept of sampling error states that samples are not perfect and we should always expect small differences between a sample mean and the population mean.

The alternative hypothesis states that there is a systematic difference between treatments that causes the difference scores to be consistently positive (or negative) and produces a non-zero mean difference between the treatments:

$$H_1: \mu_D \neq 0$$

According to the alternative hypothesis, the sample mean difference obtained in the research study is a reflection of the true mean difference that exists in the population.

The repeated-measures t statistic forms a ratio with exactly the same structure as the single-sample t statistic presented in Chapter 9. The numerator of the t statistic measures the difference between the sample mean and the hypothesized population mean. The bottom of the ratio is the standard error, which measures how much difference is reasonable to expect between a sample mean and the population mean if there is no treatment effect; that is, how much difference is expected by simply by sampling error.

$$t = \frac{\text{obtained difference}}{\text{standard error}} = \frac{M_D - \mu_D}{s_{M_D}} \qquad df = n - 1$$

For the repeated-measures t statistic, all calculations are done with the sample of difference scores. The mean for the sample appears in the numerator of the t statistic and the variance of the difference scores is used to compute the standard error in the denominator. As usual, the standard error is computed by

$$s_{M_D} = \sqrt{\frac{s^2}{n}} \quad \text{or} \quad s_{M_D} = \frac{s}{\sqrt{n}}$$

Measuring Effect Size for the Independent-Measures t

Effect size for the independent-measures t is measured in the same way that we measured effect size for the single-sample t (Chapter 9) and the independent-measures t

(Chapter 10). Specifically, you can compute an estimate of Cohen's d to obtain a standardized measure of the mean difference, or you can compute r^2 to obtain a measure of the percentage of variance accounted for by the treatment effect.

Comparing Repeated-Measures and Independent-Measures Designs

Because a repeated-measures design uses the same individuals in both treatment conditions, this type of design usually requires fewer participants than would be needed for an independent-measures design. In addition, the repeated-measures design is particularly well suited for examining changes that occur over time, such as learning or development. The primary advantage of a repeated-measures design, however, is that it reduces variance and error by removing individual differences.

The first step in the calculation of the repeated-measures t statistic is to find the difference score for each subject. This simple process has two very important consequences.

1. First, the D score for each subject provides an indication of how much difference there is between the two treatments. If all of the subjects show roughly the same D scores, then you can conclude that there appears to be a consistent, systematic difference between the two treatments. You should also note that when all the D scores are similar, the variance of the D scores will be small, which means that the standard error will be small and the t statistic is more likely to be significant.

2. Also, you should note that the process of subtracting to obtain the D scores removes the individual differences from the data. That is, the initial differences in performance from one subject to another are eliminated. Removing individual differences also tends to reduce the variance, which creates a smaller standard error and increases the likelihood of a significant t statistic.

The following data demonstrate these points.

Subject	X_1	X_2	D
A	9	16	7
B	25	28	3
C	31	36	5
D	58	61	3
E	72	79	7

First, notice that all of the subjects show an increase of roughly 5 points when they move from treatment 1 to treatment 2. Because the treatment difference is very consistent, the D scores are all clustered close together will produce a very small value for s^2. This means that the standard error in the bottom of the t statistic will be very small.

Second, notice that the original data show big differences from one subject to another. For example, subject B has scores in the 20's and subject E has scores in the 70's. However, these big **individual differences** are eliminated when the difference scores are calculated. Because the individual differences are removed, the D scores are usually much less variable than the original scores. Again, a smaller variance will produce a smaller standard error, which will increase the likelihood of a significant t statistic.

Finally, you should realize that there are potential disadvantages to using a repeated-measures design instead of independent-measures. Because the repeated-measures design requires that each individual participate in more than one treatment, there is always the risk that exposure to the first treatment will cause a change in the participants that influences their scores in the second treatment. For example, practice in the first treatment may cause improved performance in the second treatment. Thus, the scores in the second treatment may show a difference, but the difference is not caused by the second treatment. When participation in one treatment influences the scores in another treatment, the results may be distorted by **order effects**, and this can be a serious problem in repeated-measures designs.

LEARNING OBJECTIVES

1. Know the difference between independent-measures and repeated-measures research designs.

2. Know the difference between a repeated-measures and a matched-subjects research design.

3. Be able to perform the computations for the repeated-measures t test.

4. Be able to compute measures of effect size for the repeated-measures t.

5. Understand the advantages and disadvantages of the repeated-measures design (compared to an independent-measures design) and when this type of study is appropriate.

NEW TERMS AND CONCEPTS

The following terms were introduced in this chapter. You should be able to define or describe each term and, where appropriate, describe how each term is related to other terms in the list.

Repeated-measures design   A research study where the same sample of individuals is measured in all of the treatment conditions.

Matched-samples design — A research study where the individuals in one sample are matched one-to-one with the individuals in a second sample. The matching is based on a variable considered relevant to the study.

Related-samples t statistic — The single sample t statistic applied to a sample of difference scores (D values) and the corresponding population of difference scores.

Difference scores — The difference between two measurements obtained for a single subject. $D = X_2 - X_1$

Estimated standard error of M_D — An estimate of the standard distance between a sample mean difference M_D and the population mean difference μ_D

Individual differences — The naturally occurring differences from one individual to another that may cause the individuals to have different scores.

Order effects — The effects of participating in one treatment that may influence the scores in the following treatment.

NEW FORMULAS

$$D = X_2 - X_1 \qquad M_D = \frac{\Sigma D}{n}$$

$$t = \frac{M_D - \mu_D}{s_{M_D}}$$

$$s_{M_D} = \sqrt{\frac{s^2}{n}} \quad \text{or} \quad s_{M_D} = \frac{s}{\sqrt{n}}$$

$$\text{estimated Cohen's d} = \frac{M_D}{s}$$

STEP BY STEP

<u>Hypothesis Testing with the Related-Samples t</u>. The related-samples t statistic is used to test for a mean difference (μ_D) between two treatment conditions, using data from a single sample of subjects where each individual is measured first in one treatment condition and then in the second condition. This test statistic also is used for matched-subjects designs, which consists of two samples with the subjects in one sample matched one-to-one with the subjects in the second sample. Often, a repeated-measures experiment consists of a "before/after" design where each subject is measured before treatment and then again after treatment. The following example will be used to demonstrate the related-samples t test.

A researcher would like to determine whether a particular treatment has an effect on performance scores. A sample of n = 16 subjects is selected. Each subject is measured before receiving the treatment and again after treatment. The researcher records the difference between the two scores for each subject. The difference scores averaged $M_D = -6$ with SS = 960.

Step 1: State the hypotheses and select an alpha level. In the experiment the treatment was given to a sample, but the researcher wants to determine whether the treatment has any effect for the general population. As always, the null hypothesis says that there is no effect.

H_0: $\mu_D = 0$ (on average, the before/after difference for the population is zero)

The alternative hypothesis states that the treatment does produce a difference.

H_1: $\mu_D \neq 0$

We will use $\alpha = .05$.

Step 2: Locate the critical region. With a sample of n = 16, the related-samples t statistic will have df = 15. Sketch the distribution of t statistics with df = 15 and locate the extreme 5% of the distribution. The critical boundaries are t = ∀2.131.

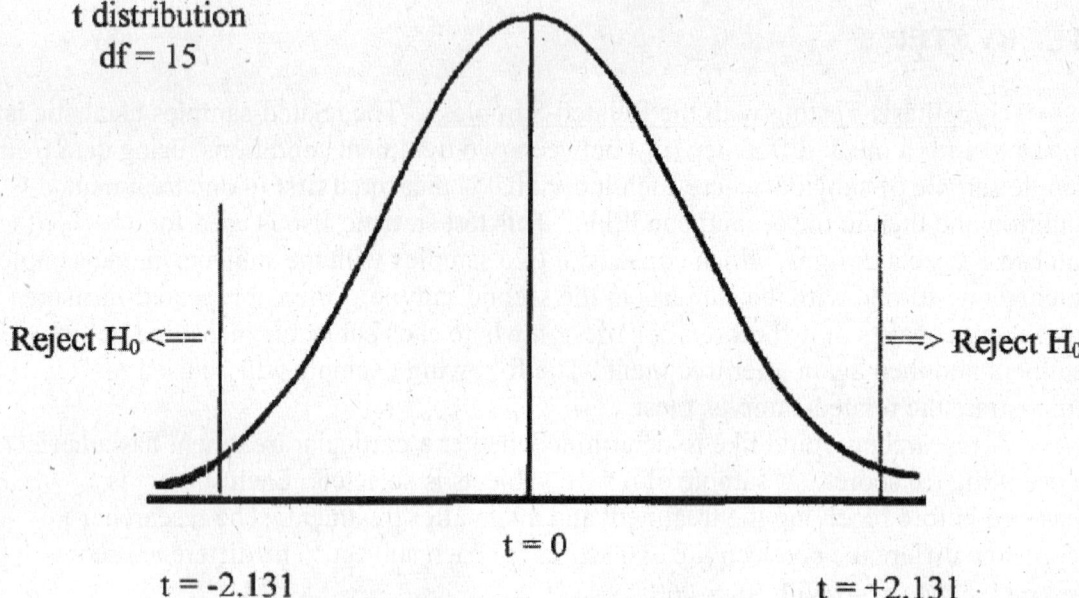

Step 3: Calculate the test statistic. As with all t statistics, it is easier to begin the calculation with the denominator of the t formula. For this example, the variance for the difference scores is

$$s^2 = \frac{SS}{n-1} = \frac{960}{15} = 64$$

With a sample of n = 16, the standard error is

$$s_{M_D} = \sqrt{\frac{s^2}{n}} = \sqrt{\frac{64}{16}} = \frac{8}{4} = 2$$

Finally, substitute the sample mean difference and the standard error in the t formula,

$$t = \frac{M_D - \mu_D}{s_{M_D}} = \frac{-6 - 0}{2} = \frac{-6}{2} = -3.00$$

Step 4: Make Decision. The t statistic is in the critical region. This is a very unlikely value for t if H_0 is true. Therefore, we reject H_0. The researcher concludes that the treatment does have a significant effect on performance scores.

Measuring Effect Size for the Repeated-Measures t

We will use the data from the preceding example to demonstrate the calculation of Cohen's d and r^2. (Note that the sign of the mean difference is usually not included in the calculation of Cohen's d.)

$$\text{estimated Cohen's d} = \frac{M_D}{s} = \frac{6}{\sqrt{64}} = \frac{6}{8} = 0.75$$

$$r^2 = \frac{t^2}{t^2 + df} = \frac{(-3.00)^2}{(-3.00)^2 + 15} = \frac{9}{24} = 0.375$$

HINTS AND CAUTIONS

1. It is important to remember that the related-samples analysis is based on difference scores (D scores). Therefore, the computations of s^2 and s_{M_D} are based on the sample of D scores.
2. When calculating the difference scores, be sure all of them are obtained by subtracting in the same direction. That is, you may either subtract 1st – 2nd, or 2nd – 1st as long as the same method is used throughout.

SELF TEST

True/False Questions

1. A researcher would like to compare two treatment conditions with a set of 30 scores in each treatment. If a repeated-measures design is used, the study will require only n = 60 participants.

2. Although a repeated-measures study measures two scores for each participant, the sample mean and variance are computed using only one score for each participant.

3. In a repeated measures study, the null hypothesis says that the mean for the sample of difference scores should be equal to zero.

4. In the repeated-measures t statistic, the value for the estimated standard error in the denominator is computed entirely from the sample data.

5. In a repeated-measures study comparing two treatments with a sample of n = 15 participants, the researcher measures two scores for each individual to obtain a total of 30 scores. The repeated-measures t statistic for this study has df = 29.

6. If a set of n = 9 difference scores has a mean of $M_D = 3.5$ and a variance of $s^2 = 36$, then the standard error for the sample mean difference is 4 points.

7. A set of n = 16 difference scores has a mean of $M_D = 4$ and a variance of $s^2 = 36$. Cohen's d for this sample is d = 4/6.

8. A repeated-measures study with a sample of n = 16 participants produces a repeated-measures t = 2.00. If effect size is measured using r^2, then $r^2 = 4/20$.

9. High variance for a sample of difference scores indicates that the treatment does not have a consistent effect.

10. If other factors are held constant, the higher the variance is for a sample of difference scores, the lower the likelihood of rejecting the null hypothesis.

Multiple Choice Questions

1. A repeated-measures study would *not* be appropriate for which of the following situations?
    a. A researcher would like to study the effect of practice on performance.
    b. A researcher would like to compare individuals from two different populations.
    c. The effect of a treatment is studied in a small group of individuals with a rare disease.
    d. A developmental psychologist examines how behavior unfolds by observing the same group of children at different ages.

2. The following data were obtained from a repeated-measures research study. What is the value for M_D?

    a. $M_D = 0$
    b. $M_D = 2$
    c. $M_D = 2.5$
    d. $M_D = 8$

Subject	1st	2nd
#1	8	10
#2	6	12
#3	10	7
#4	9	11
#5	7	10

3. A repeated-measures study comparing two treatments measures two scores for each individual in a sample of n = 15 participants. What is the value of df for the t statistic?
   a. 14
   b. 15
   c. 29
   d. 30

4. A repeated-measures study and a matched-subjects study are both used to compare two treatments. If each study uses a total of 30 participants, then what are the df values for the two studies?
   a. repeated-measures df = 29 and matched-subjects df = 29
   b. repeated-measures df = 29 and matched-subjects df = 14
   c. repeated-measures df = 14 and matched-subjects df = 29
   d. repeated-measures df = 14 and matched-subjects df = 29

5. A repeated-measures using a sample of n = 20 participants would produce a t statistic with df = ____.
   a. 9
   b. 19
   c. 20
   d. 39

6. A sample of n = 4 difference scores produces SS = 48. What is the estimated standard error for the sample mean difference?
   a. 12
   b. 16
   c. 4
   d. 2

7. A repeated-measures experiment and an independent-measures experiment both produce t statistics with df = 20. How many individuals participated in each experiment?
   a. n = 21 for independent-measures and n = 21 for repeated measures
   b. n = 21 for independent-measures and n = 11 for repeated measures
   c. n = 22 for independent-measures and n = 21 for repeated measures
   d. n = 22 for independent-measures and n = 11 for repeated measures

8. In general, what characteristics of the difference scores are most likely to produce a significant t statistic for the repeated-measures hypothesis test?
   a. a large number of scores and a large variance
   b. a large number of scores and a small variance
   c. a small number of scores and a large variance
   d. a small number of scores and a small variance

9. What is indicated by a large variance for a sample of difference scores?
   a. a consistent treatment effect and a high likelihood of a significant difference
   b. a consistent treatment effect and a low likelihood of a significant difference
   c. an inconsistent treatment effect and a high likelihood of a significant difference
   d. an inconsistent treatment effect and a low likelihood of a significant difference

10. A sample of difference scores has a mean of $M_D = 5$. Which combination of factors is most likely to produce a significant difference?
    a. $n = 4$ with a variance of $s^2 = 36$ for the difference scores
    b. $n = 4$ with a variance of $s^2 = 64$ for the difference scores
    c. $n = 9$ with a variance of $s^2 = 36$ for the difference scores
    d. $n = 9$ with a variance of $s^2 = 64$ for the difference scores

11. A researcher is using a repeated-measures study to evaluate the difference between two treatments. If there is a consistent difference between the treatments then the data should produce _____.
    a. a small variance for the difference scores and a small standard error
    b. a small variance for the difference scores and a large standard error
    c. a large variance for the difference scores and a small standard error
    d. a large variance for the difference scores and a large standard error

12. For which of the following situations would a repeated-measures design have a substantial advantage over an independent-measures design.
    a. few subjects are available and individual differences are large
    b. few subjects are available and individual differences are small
    c. many subjects are available and individual differences are large
    d. many subjects are available and individual differences are small

13. A sample of difference scores has a mean of $M_D = 5$ with a variance of $s^2 = 100$. If effect size is measured using Cohen's d, what is the value of d?
    a. d = 5/10
    b. d = 5/100
    c. cannot determine without knowing the sample size

14. Which of the following would have little or no influence on effect size as measured by Cohen's d or by r^2?
    a. Increasing the sample size
    b. Increasing the sample mean difference
    c. Increasing the sample variance
    d. All of the other three options would influence the magnitude of effect size.

15. Which of the following is *not true* of a repeated-measures design compared to an independent-measures design?
    a. The repeated-measures design requires fewer participants.
    b. The repeated-measures design tends to produce a smaller standard error.
    c. The repeated-measures design is less likely to reject the null hypothesis.
    d. The repeated-measures design has no risk that the participants in one treatment are noticeably different from the participants in the other treatment.

Other Questions

1. A researcher would like to conduct a study comparing two treatment conditions with 30 individuals measured in each treatment.
    a. How many participants are needed fdor an independent-measures design?
    b. How many participants are needed for a repeated-measures design?
    c. How many participants are needed for a matched-subjects design?

2. The following data represent the results from a repeated-measures study comparing two treatment conditions.

Participant	Treatment 1st	2nd	D
#1	8	14	
#2	6	11	
#3	10	10	
#4	9	11	
#5	7	12	
#6	10	16	

    a. Calculate the difference scores and compute the mean difference, M_D, and the variance for the sample of difference scores.
    b. Do the results indicate a significant difference between the two treatments? Use a two-tailed test with $\alpha = .05$.
    c. Calculate r^2 and the estimated Cohen's d to measure the size of the treatment effect.

3. A researcher would like to test the effect of a new diet drug on the activity level of animals. A sample of $n = 16$ rats is obtained and each rat's activity level is measured on an exercise wheel for one hour prior to receiving the drug. Thirty minutes after receiving the drug, each rat is again tested on the activity wheel. The data show that the rats increased their activity by an average of $M_D = 21$ revolutions with $SS = 6000$ after receiving the drug. Do these data indicate that the drug had a significant effect on activity. Test at the .05 level.
    a. Using symbols, state the hypotheses for this test.
    b. Locate the critical region for $\alpha = .05$.
    c. Compute the t statistic for these data.
    d. What decision should the researcher make?
    e. Compute the effect size for this study using r^2 (the percentage of variance accounted for).

ANSWERS TO SELF TEST

True/False Answers

1. False. The study requires only 30 participants.
2. True
3. False. The null hypothesis says that the population mean difference is zero.
4. True
5. False. The 30 scores are used to compute 15 difference scores: df = 14.
6. False. The standard error is $\sqrt{4}$
7. True
8. False. $r^2 = 4/19$
9. True
10. True

Multiple-Choice Answers

1. b Two different populations will require two different samples.
2. b The sum of the difference score is 10 and the mean is 10/5 = 2
3. a df = n − 1.
4. b The matched-subjects study uses 15 matched pairs.
5. b df = n − 1
6. d The variance is 16 and the standard error is the square root of 16/4.
7. c The independent-measures study requires 22 participants (probably two groups, each with n = 11) and the repeated-measures study has only one group of n = 21.
8. b The largest sample size and the smallest variance produce the smallest standard error.
9. d A large variance indicates that the difference scores are widely scattered.
10. c The largest sample size and the smallest variance produce the smallest standard error.
11. a Consistent difference scores produce a smaller variance and less error.
12. a The repeated-measures design requires few participants and removes the individual differences.
13. a Cohen's d = the sample mean difference divided by the standard deviation.
14. a Sample size has little or no influence on measures of effect size.
15. c A repeated-measures design typically has a smaller standard error and is more likely to detect a real treatment effect.

Other Answers

1. a. An independent-measures design would require two separate samples of n = 30 for a total of 60 subjects.
   b. A repeated-measures design would require only one sample of n = 30 subjects.
   c. A matched-subjects design would require two matched samples of n = 30 for a total of 60 subjects.

2. a.

Participant	Treatment 1st	2nd	D	
#1	8	14	+6	$\Sigma D = 24$
#2	6	11	+5	
#3	10	10	0	$M_D = 24/6 = 4$
#4	9	11	+2	
#5	7	12	+5	$s^2 = 30/5 = 6$
#6	10	16	+6	

   b. The estimated standard error is 1, and $t = 4/1 = 4.00$. With $df = 5$, the critical value is $t = \pm 2.571$. Reject the null hypothesis and conclude that there is a significant difference between the two treatments.
   c. estimated Cohen's $d = 4/\sqrt{6} = 1.63$ and $r = 16/(16 + 5) = 0.762$

3. a. $H_0: \mu_D = 0$ (no effect). $H_1: \mu_D \neq 0$.
   b. With $df = 15$ and $\alpha = .05$, the critical value are $t = \pm 2.131$.
   c. The sample variance is $s^2 = 400$, the standard error is 5 points, and $t(15) = 4.2$.
   d. Reject the null hypothesis. The diet drug has a significant effect on activity.
   e. For these data, $r^2 = 17.64/32.64 = 0.54$ (or 54%)

CHAPTER 12

ESTIMATION

CHAPTER SUMMARY

The goals of Chapter 12 are:
1. To introduce the statistical technique of estimation as an alternative to a hypothesis test.
2. To demonstrate point and interval estimates for the single-sample t, the independent-measures t, and the repeated-measures t.
3. To demonstrate how the width of a confidence interval is related to the sample size and the level of confidence.

Estimation

In general terms, **estimation** uses a sample statistic as the basis for estimating the value of the corresponding population parameter. Although estimation and hypothesis testing are similar in many respects, they are complementary inferential processes. A hypothesis test is used to determine whether or not a treatment has an effect, while estimation is used to determine how much effect.

This complementary nature is demonstrated when estimation is used after a hypothesis test that resulted in rejecting the null hypothesis. In this situation, the hypothesis test has established that a treatment effect exists and the next logical step is to determine how much effect. It is also common to use estimation in situations where a researcher simply wants to learn about an unknown population. In this case, a sample is selected from the population and the sample data are then used to estimate the population parameters.

You should keep in mind that even though estimation and hypothesis testing are inferential procedures, these two techniques differ in terms of the type of question they address. A hypothesis test, for example, addresses the somewhat academic question concerning the *existence* of a treatment effect. Estimation, on the other hand, is directed toward the more practical question of *how much* effect.

The estimation process can produce either a **point estimate** or an **interval estimate**. A point estimate is a single value and has the advantage of being very precise. For example, based on sample data, you might estimate that the mean age for students at the state college is $\mu = 21.5$ years. An interval estimate consists of a range of values and has the advantage of providing greater confidence than a point estimate. For example, you might estimate that the mean age for students is somewhere between 20 and 23 years. Note that the interval estimate is less precise, but gives more confidence. For this reason, interval estimates are usually called **confidence intervals**.

Estimation with the Three t Statistics

There is an estimation procedure that accompanies each of the three t tests presented in the preceding three chapters. The estimation process begins with the same t statistic that is used for the corresponding hypothesis test. These all have the same conceptual structure:

$$t = \frac{\text{sample statistic - unknown parameter}}{\text{standard error}}$$

The basic estimation equation is obtained by solving this equation for the unknown parameter:

$$\text{unknown parameter} = \text{statistic} \ \forall \ (t)(\text{standard error})$$

For the single sample t, the unknown parameter is the population mean, μ. For the independent-measures t, the parameter is the difference between population means, $\mu_1 - \mu_2$, and for the repeated-measures t, the unknown parameter is the population mean difference, μ_D. To use the estimation equation, you first must obtain a value for t by estimating the location of the data within the appropriate t distribution. For a point estimate, a value of zero is used for t, yielding a single value for your estimate of the unknown population parameter. For interval estimates, you first select a level of confidence and then find the corresponding interval range of t values in the t distribution table. The estimated value for t (or range of values) is then substituted in the equation along with the values for the sample statistic and the standard error. The equation is then solved for the unknown parameter.

Factors Influencing the Width of a Confidence Interval

If other factors are held constant, increasing the level of confidence (for example, from 80% to 90%), will cause in increase in the width of the confidence interval. To obtain greater confidence, you must use a wider range of t statistic values, which results in a wider interval. On the other hand, increasing the sample size will cause a decrease in the width of the confidence interval. In simple terms, a larger sample gives you more information, which means that you can estimate the population parameter with more precision.

LEARNING OBJECTIVES

1. You should be able to use sample data to make a point estimate or an interval estimate of an unknown population mean using a single sample t statistic.

2. You should be able to make a point estimate or an interval estimate of a population mean difference using sample data from an independent-measures or a repeated-measures study.

3. You should understand how the size of a sample influences the width of a confidence interval.

4. You should understand how the level of confidence (the % confidence) influences the width of a confidence interval.

NEW TERMS AND CONCEPTS

The following terms were introduced in this chapter. You should be able to define or describe each term and, where appropriate, describe how each term is related to other terms in the list.

Estimation — The inferential process of using a sample statistic to estimate the value of an unknown population parameter.

Point estimate — A single number is used to estimate the value of an unknown parameter. The result is a precise estimate, but without confidence.

Interval estimate — A range of values is used to estimate an unknown parameter. The result is a less precise estimate but there is a gain in confidence.

Confidence interval — An interval estimate that is described in terms of the level (percentage) of confidence in the accuracy of the estimation.

NEW FORMULAS

$\mu = M \ \forall \ ts_M$

$\mu_D = M_D \ \forall \ ts_{M_D}$

$(\mu_1 - \mu_2) = (M_1 - M_2) \ \forall \ ts_{M_1 - M_2}$

STEP BY STEP

<u>Estimation with the t statistic</u>: The following example will be used to demonstrate the step-by-step procedures for estimation with the t statistic.

A researcher begins with a normal population with $\mu = 60$. The researcher is evaluating a specific treatment that is expected to increase scores. The treatment is administered to a sample of $n = 16$ individuals, and the mean for the treated sample is $M = 66$ with $SS = 1215$. The goal is to estimate the size of the treatment effect.

Step 1: Begin with the basic formula for estimation. Again, this is simply the regular t formula that has been solved for μ. For the single-sample t statistic,

$$\mu = M \mp ts_M$$

Step 2: Determine whether you are computing a point estimate or an interval estimate. If you want an interval, you must specify a level of confidence. We will compute a point estimate and a 90% confidence interval for the unknown population mean.

Step 3: Find the appropriate t values to substitute in the equation. For a point estimate, always use t = 0 which is the point in the exact middle of the distribution. For a 90% confidence interval, we want the t values that form the boundaries for the middle 90% of the distribution. These values are obtained from the t distribution table using df = n - 1 = 15. You should find that 90% of the t statistics with df = 15 are located between t = +1.753 and t = −1.753.

Step 4: Compute M and s_M from the sample data. For this example we are given M = 66 and you can compute

$$s^2 = \frac{SS}{n-1} = \frac{1215}{15} = 81$$

$$s_M = \sqrt{\frac{s^2}{n}} = \sqrt{\frac{81}{16}} = \frac{9}{4} = 2.25$$

Step 5: Substitute the appropriate values in the estimation equation.

point estimate	interval estimate
μ = 66 ∓ 0	μ = 66 ∓ (1.753)(2.25)
μ = 66	μ = 66 ∓ 3.94
Estimate μ = 66	Estimate μ between 62.06 and 69.94

HINTS AND CAUTIONS

1. When computing a confidence interval after a hypothesis test, many students incorrectly take the t value that was computed in the hypothesis test and use this value in the estimation equation. Remember, the t value in the estimation

equation is determined by the level of confidence and is obtained from the t distribution table.

2. When trying to locate the appropriate t value for a confidence interval, remember to use the "proportion in two tails" column of the t table. For an 80% confidence interval, for example, there would be 20% of the distribution left in the two tails and you should use the .20 column for two tails of the distribution.

SELF TEST

True/False Questions

1. The advantage of a point estimate compared to an interval estimate is that the point estimate provides more confidence.

2. Estimation is commonly used after a hypothesis test when the decision is to fail to reject the null hypothesis.

3. Point estimates of the population mean or mean difference always use t = 0 in the estimation equation.

4. For an 80% confidence interval we simply estimate that the sample produces a t statistic somewhere in the middle 80% of the t distribution and then use the t values corresponding to the middle 80% to compute the interval.

5. If a sample of n = 9 scores is used to make a 90% confidence interval estimate of the population mean, then values of t = ±2.306 would be used in the equation.

6. A researcher conducts a study to determine whether a new drug is effective for treating high blood pressure. The researcher should use a hypothesis test, rather than estimation, to evaluate the data.

7. If other factors are held constant, increasing the sample size from n = 10 to n = 20 will increase the width of a confidence interval.

8. If other factors are held constant, increasing the percentage of confidence from 80% to 90% will increase the width of a confidence interval.

9. If the results from an independent-measures experiment lead to rejecting the null hypothesis with α = .05, then the value $\mu_1 - \mu_2 = 0$ will be contained in the 95% confidence interval for the population mean difference.

10. A sample in treatment I has a mean of M = 12 with SS = 23. A second sample is given treatment II and has M = 16 with SS = 29. If these data are used to estimate the population mean difference between the two treatments, then the point estimate would be 4 points.

Multiple-Choice Questions

1. Compared to a point estimate, an interval estimate _____.
    a. has greater precision and greater confidence
    b. has greater precision but less confidence
    c. has less precision but greater confidence
    d. has less precision and less confidence

2. A sample of n = 5 scores has a mean of M = 45 and a variance of s^2 = 20. Using this sample, the 90% confidence interval for the population mean is
    a. $\mu = 45 \pm 2.776(\sqrt{20})$
    b. $\mu = 45 \pm 2.776(2)$
    c. $\mu = 45 \pm 2.132(\sqrt{20})$
    d. $\mu = 45 \pm 2.132(2)$

3. The wider the confidence interval, the less precise the estimate is. With this in mind, which combination of factors will produce the most precise estimate of the population mean?
    a. a sample of n = 20 with 80% confidence
    b. a sample of n = 20 with 90% confidence
    c. a sample of n = 50 with 80% confidence
    d. a sample of n = 50 with 90% confidence

4. If other factors are held constant, which combination of factors will produce the widest 90% confidence interval for the population mean?
    a. a sample of n = 20 with s^2 = 40
    b. a sample of n = 50 with s^2 = 40
    c. a sample of n = 20 with s^2 = 10
    d. a sample of n = 50 with s^2 = 10

5. A researcher would like to determine how much difference there is in the vocabulary skills for 3-year-old girls compared with 3-year-old boys. If the researcher constructs an 80% confidence interval for the mean difference using a sample of n = 10 boys and n = 10 girls, what t values should be used in the estimation equation?
    a. ±1.330
    b. ±1.328
    c. ±1.734
    d. ±1.729

6. A researcher conducts a repeated-measures study to determine how much difference exists between two treatment conditions. If the 95% confidence interval extends from 8.00 to 10.00, then what is the value of the sample mean difference?
    a. 2
    b. 9
    c. 18
    d. cannot be determined without more information

7. The results from a repeated-measures experiment produce a 95% confidence interval estimate of µ that extends from –2.00 to +14.00. If a hypothesis test is conducted with the same data, what decision would be reached?
    a. reject the null hypothesis with α = .05
    b. reject the null hypothesis with α = .05 or α = .01
    c. fail to reject the null hypothesis with α = .05
    d. fail to reject the null hypothesis with α = .05 or α = .01

8. A researcher obtains a sample of n = 25 individuals from a population with unknown parameters. The goal is to use the sample to estimate the population mean. If the sample mean is M = 84 and SS = 2400, then what value should be used as the point estimate of µ?
    a. 0
    b. 25
    c. 84
    d. cannot answer without additional information

9. Which combination of factors would definitely *increase* the width of a confidence interval?
    a. use a larger sample and increase the level of confidence
    b. use a smaller sample and increase the level of confidence
    c. use a larger sample and decrease the level of confidence
    d. use a smaller sample and decrease the level of confidence

10. A researcher obtains a sample of n = 25 individuals from a population with unknown parameters. The goal is to use the sample to estimate the population mean. If the sample mean is M = 84 and SS = 2400, then the 80% confidence interval would be _____.
    a. $\mu = 84 \pm 10(1.318)$
    b. $\mu = 84 \pm 2(1.318)$
    c. $\mu = 84 \pm 10(1.711)$
    d. $\mu = 84 \pm 2(1.711)$

11. A researcher obtains a sample of n = 9 individuals in tests each person in two different treatment conditions. The sample mean difference between treatments is M_D = 12 points with SS = 72. Which of the following is the correct equation for the 80% confidence interval for the population mean difference?
    a. $\mu_D = 0 \pm 3(1.397)$
    b. $\mu_D = 0 \pm 1(1.397)$
    c. $\mu_D = 12 \pm 3(1.397)$
    d. $\mu_D = 12 \pm 1(1.397)$

12. A researcher constructs a 90% confidence interval for the population mean using a sample of n = 25 individuals, but is disappointed because the estimate is not very precise because the interval is too wide. What can be done to produce a narrower interval?
    a. Use a larger sample.
    b. Switch to 80% confidence.
    c. Both a and b will produce a narrower interval.
    d. Neither a nor b will produce a narrower interval.

13. A researcher would like to determine how the average mathematical skill level for freshmen entering the college compares with the average from freshmen 10 years ago. A sample of n = 30 freshmen is obtained and each student is given a standardized mathematics skills test. The researcher plans to use the sample data to estimate the mean score for the current population of freshmen and then compare the result with the mean obtained 10 years ago. Which estimation equation should the researcher use?
    a. the single-sample t equation
    b. the independent-measures t equation
    c. the repeated-measures t equation
    d. cannot determine without additional information

14. A researcher knows that a special reading program will improve students' reading scores. To determine how much improvement will occur, the researcher obtains a sample of n = 25 students and measures each individual's reading performance before and after the special program. The data will be used to estimate the mean amount of improvement. Which estimation equation should the researcher use?
    a. the single-sample t equation
    b. the independent-measures t equation
    c. the repeated-measures t equation
    d. cannot determine without additional information

15. The 90% confidence interval for the difference between two population means extends from 4.00 to 8.00. Based on this information you can conclude that the difference between the two sample means was _____.
    a. 2 points
    b. 4 points
    c. 6 points
    d. 8 points

Other Questions

1. Estimation often is used after a hypothesis test. Although these two inferential techniques involve many of the same calculations, they are intended to answer different questions.
    a. In general terms, what information is provided by a hypothesis test and what information is provided by estimation?
    b. Explain why it would not be appropriate to use estimation after a hypothesis test where the decision was "fail to reject" the null hypothesis.

2. A researcher would like to evaluate the reading ability of fourth-grade students in the city school district. A sample of n = 36 students is obtained and the students are given a standardized reading test for which the nationwide fourth-grade average is μ = 40. The students in the sample averaged M = 42.5 with SS = 2835.
   a. Make a point estimate of the mean reading score for the entire population of city school fourth-grade students. How do they compare with the national average?
   b. Make an interval estimate of the population mean so that you are 90% confident that the new mean is in your interval. Again, how do the city school students compare with the national average?

3. A researcher would like to evaluate the effect of a regular exercise program on the health and well being of elderly adults. A sample of 16 participants is obtained from a local senior center and each person completes a life-satisfaction questionnaire. Then the participants are enrolled in an exercise class that meets twice weekly for one hour. After six weeks, each participant again completes the life-satisfaction questionnaire. On average, the participants scored M_D = 8 points higher after the program with SS = 960.
   a. On the basis of these data, can the counselor conclude that the exercise program has a significant effect on well being? Test at the .05 level of significance.
   b. Use the data to make a point estimate of how much improvement results from regular exercise.
   c. Make an interval estimate of the mean improvement so that you are 90% confident that the true mean is contained in your interval.

4. The government has developed a pamphlet listing driving tips that are designed to promote fuel economy. To test the effectiveness of these tips, a sample of twenty families is obtained, all with identical cars and similar driving habits. Ten of these families are given the pamphlet and instructions to follow the driving tips carefully. The other ten families are simply instructed to monitor their gas mileage for a two-week period. At the end of two weeks, the gas mileage figures for both groups are as follows:

Experimental Group with Pamphlet	Control Group no Pamphlet
n = 10	n = 10
M = 25.7	M = 21.6
SS = 160	SS = 128

   a. Do these data indicate that the driving tips have a significant effect on gas mileage? Test at the .05 level of significance.
   b. Use the data to make a point estimate of how much improvement in gas mileage results from following the tips.
   c. Use the data to construct a 90% confidence interval to estimate how much effect the driving tips have on gas mileage.

ANSWERS TO SELF TEST

True/False Answers

1. False. A point estimate is more precise but gives less confidence.
2. False. Failing to reject the null hypothesis means that you have concluded that the treatment has no effect. In this case, it would make no sense to estimate "how much" effect.
3. True
4. True
5. False. With df = 8, the 90% t values are ±1.860.
6. True
7. False. A larger sample produces a smaller error and a narrower interval.
8. True
9. True
10. True

Multiple-Choice Answers

1. c Point estimates are precise but have no confidence.
2. d With df = 4 and 90% confidence, t = ±2.132 and the standard error is 2.
3. c A large sample and low confidence combine for the narrowest interval.
4. a A large sample and small variance produce a large standard error and a wider interval.
5. a For an independent-measures study, df = 18.
6. b The sample mean is in the middle of the confidence interval
7. a A mean difference of zero is an acceptable value with 95% confidence. Definitely reject the null hypothesis with $\alpha = .05$.
8. c Always use the sample mean as the point estimate of μ.
9. b High confidence and a small sample combine to produce a wide interval.
10. b The sample has df = 24 and an estimated standard error of 2 points.
11. d The interval is built around the sample mean and the standard error is 1.
12. c A larger sample and lower confidence will both reduce the width.
13. a The study has one sample mean being compared with a preexisting mean.
14. c The same students are measured twice. This is a repeated-measures study.
15. c The sample mean (or mean difference) is always in the center of the confidence interval.

Other Answers

1. a. A hypothesis test is intended to determine whether or not a treatment effect exists. Estimation is used to determine how much effect.
   b. If the decision from the hypothesis test is "fail to reject H_0" then the data do not provide sufficient evidence to conclude that there is any treatment effect. In this case, it would not be reasonable to use estimation in an attempt to determine "how much" effect exists.

2. a. Use the sample mean, M = 42.5 as the point estimate of μ. The students appear to be slightly above the national average.
    b. Using df = 30 because 35 is not in the table, the 90% confidence interval would be μ = 42.5 ∀ (1.697)(1.5). The interval extends from 39.95 to 45.05. Again, the students appear to be somewhat above average, however there may be no difference at all because the national mean is contained within the interval.

3. a. With s^2 = 64 and s_{M_D} = 2, the data have a t statistic of t = 4.00. Reject H_0 and conclude that the exercise has a significant effect.
    b. Use the sample mean, M_D = 8 points, as the point estimate of μ_D.
    c. The 90% confidence interval would be μ_D = 8 ∀ (1.753)(2), which gives a mean improvement between 4.494 to 11.506.

4. a. The pooled variance is 16 and the estimated standard error is 1.79. The t statistic is t = 2.29. Reject H_0.
    b. The sample mean difference, 4.1 miles per gallon, is the best point estimate.
    c. The 90% confidence interval would be $\mu_1 - \mu_2$ = 4.1 ∀ (1.734)(1.79), which gives an interval estimate between 1.00 and 7.2 miles per gallon.

CHAPTER 13

INTRODUCTION TO ANALYSIS OF VARIANCE

CHAPTER SUMMARY

The goals of Chapter 13 are:
1. To introduce the logical foundation for analysis of variance.
2. To demonstrate the process of hypothesis testing with analysis of variance.
3. To demonstrate the use of post tests to help interpret the results from an analysis of variance.
4. To demonstrate the measurement of effect size for an analysis of variance.
5. To demonstrate the relationship between analysis of variance and the independent-measures t test for a study comparing exactly two samples or treatment conditions.

The Logic and the Process of Analysis of Variance
   Chapter 13 presents the general logic and basic formulas for the hypothesis testing procedure known as **analysis of variance** (ANOVA). The purpose of ANOVA is much the same as the t tests presented in the preceding three chapters: the goal is to determine whether the mean differences that are obtained for sample data are sufficiently large to justify a conclusion that there are mean differences between the populations from which the samples were obtained. The difference between ANOVA and the t tests is that ANOVA can be used in situations where there are *two or more* means being compared, whereas the t tests are limited to situations where only two means are involved.
   Analysis of variance is necessary to protect researchers from excessive risk of a Type I error in situations where a study is comparing more than two population means. These situations would require a series of several t tests to evaluate all of the mean differences. (Remember, a t test can compare only 2 means at a time.) Although each t test can be done with a specific α-level (risk of Type I error), the α-levels accumulate over a series of tests so that the final **experimentwise α-level** can be quite large. ANOVA allows researcher to evaluate all of the mean differences in a single hypothesis test using a single α-level and, thereby, keeps the risk of a Type I error under control no matter how many different means are being compared.
   Although ANOVA can be used in a variety of different research situations, this chapter presents only independent-measures designs involving only one independent variable.
   The test statistic for ANOVA is an F-ratio, which is a ratio of two sample variances. In the context of ANOVA, the sample variances are called **mean squares**, or **MS** values. The top of the F-ratio $MS_{between}$ measures the size of mean differences between samples. The bottom of the ratio MS_{within} measures the magnitude of differences that would be expected without any treatment effects.
   Thus, the F-ratio has the same basic structure as the independent-measures t statistic presented in Chapter 10.

$$F = \frac{\text{obtained mean differences (including treatment effects)}}{\text{differences expected by chance (without treatment effects)}} = \frac{MS_{between}}{MS_{within}}$$

A large value for the F-ratio indicates that the obtained sample mean differences are greater than would be expected if the treatments had no effect.

Each of the sample variances, MS values, in the F-ratio is computed using the basic formula for sample variance:

$$\text{sample variance} = MS = \frac{SS}{df}$$

To obtain the SS and df values, you must go through an analysis that separates the total variability for the entire set of data into two basic components: between-treatment variability (which will become the numerator of the F-ratio), and within-treatment variability (which will be the denominator). The two components of the F-ratio can be described as follows:

Between-Treatments Variability: $MS_{between}$ measures the size of the differences between the sample means. For example, suppose that three treatments, each with a sample of $n = 5$ subjects, have means of $M_1 = 1$, $M_2 = 2$, and $M_3 = 3$. Notice that the three means are different; that is, they are variable. By computing the variance for the three means we can measure the size of the differences. Although it is possible to compute a variance for the set of sample means, it usually is easier to use the total, T, for each sample instead of the mean, and compute variance for the set of T values. Logically, the differences (or variance) between means can be caused by two sources:
1. Treatment Effects: If the treatments have different effects, this could cause the mean for one treatment to be higher (or lower) than the mean for another treatment.
2. Chance or Sampling Error: If there is no treatment effect at all, you would still expect some differences between samples. Mean differences from one sample to another are an example of random, unsystematic *sampling error*, a concept that was first introduced in Figure 1.2 in your textbook.

Within-Treatments Variability: MS_{within} measures the size of the differences that exist *inside* each of the samples. Because all the individuals in a sample receive exactly the same treatment, any differences (or variance) within a sample cannot be caused by different treatments. Thus, these differences are caused by only one source:
1. Chance or Error: The unpredictable differences that exist between individual scores are not caused by any systematic factors and are simply considered to be random chance or error.

Considering these sources of variability, the structure of the F-ratio becomes,

$$F = \frac{\text{treatment effect} + \text{chance/error}}{\text{chance/error}}$$

When the null hypothesis is true and there are no differences between treatments, the F-ratio is balanced. That is, when the "treatment effect" is zero, the top and bottom of the F-ratio are measuring the same variance. In this case, you should expect an F-ratio near 1.00. When the sample data produce an F-ratio near 1.00, we will conclude that there is no significant treatment effect.

On the other hand, a large treatment effect will produce a large value for the F-ratio. Thus, when the sample data produce a large F-ratio we will reject the null hypothesis and conclude that there are significant differences between treatments.

To determine whether an F-ratio is large enough to be significant, you must select an α-level, find the df values for the numerator and denominator of the F-ratio, and consult the F-distribution table to find the critical value.

Analysis of Variance and Post Tests

The null hypothesis for ANOVA states that for the general population there are no mean differences among the treatments being compared; $H_0: \mu_1 = \mu_2 = \mu_3 = \ldots$
When the null hypothesis is rejected, the conclusion is that there are significant mean differences. However, the ANOVA simply establishes that differences exist, it does not indicate exactly which treatments are different. With more than two treatments, this creates a problem. Specifically, you must follow the ANOVA with additional tests, called **post tests**, to determine exactly which treatments are different and which are not. The **Scheffe test** and **Tukey's HSD** are examples of post tests. These tests are done after an ANOVA where H_0 is rejected with more than two treatment conditions. The tests compare the treatments, two at a time, to test the significance of the mean differences.

Measuring Effect Size for an Analysis of Variance

As with other hypothesis tests, an ANOVA evaluates the *significance* of the sample mean differences; that is, are the differences bigger than would be reasonable to expect just by chance. With large samples, however, it is possible for relatively small mean differences to be statistically significant. Thus, the hypothesis test does not necessarily provide information about the actual size of the mean differences. To supplement the hypothesis test, it is recommended that you calculate a measure of effect size. For an analysis of variance the common technique for measuring effect size is to compute the percentage of variance that is accounted for by the treatment effects. For the t statistics, this percentage was identified as r^2, but in the context of ANOVA the percentage is identified as η^2 (the Greek letter eta, squared). The formula for computing effect size is:

$$\eta^2 = \frac{SS_{\text{between treatments}}}{SS_{\text{total}}}$$

LEARNING OBJECTIVES

1. You should be familiar with the purpose, terminology, and special notation of analysis of variance.

2. You should be able to perform an analysis of variance for the data from a single-factor, independent-measures experiment.

3. You should recognize when post hoc tests are necessary and you should be able to complete an analysis of variance using Tukey's HSD or the Scheffe post hoc test.

4. You should be able to report the results of an analysis of variance using either a summary table or an F-ratio (including df values). Also, you should be able to understand and interpret these reports when they appear in scientific literature.

5. You should be able to compute η^2 (the percentage of variance accounted for) as a measure of effect size for an analysis of variance.

NEW TERMS AND CONCEPTS

The following terms were introduced in this chapter. You should be able to define or describe each term and, where appropriate, describe how each term is related to other terms in the list.

Factor	In analysis of variance, an independent variable (or quasi-independent variable) is called a factor.
Levels of a factor	The specific conditions or values that are used to represent the factor are called levels.
F-ratio	The test statistic for analysis of variance is called an F-ratio and compares the differences (variance) between treatments with the differences (variance) that are expected by chance.
MS (Mean Square)	In analysis of variance, a sample variance is called a mean square or MS, indicating that variance measures the mean of the squared deviations.
Post hoc test	A test that is conducted after an ANOVA with more than two treatment conditions where the null hypothesis was rejected. The purpose of post hoc tests is to determine exactly which treatment conditions are significantly different.

Between-treatments SS, df, MS   Values used to measure and describe the differences between treatments (mean differences).

Within-treatments SS, df, MS   Values used to measure and describe the differences inside the treatment conditions. These differences are assumed to measure chance or error variability.

Total SS and df   Values used to measure and describe the total amount of variability for the entire set of data.

η^2 (eta squared)   A measure of effect size based on the percentage of variance accounted for by the sample mean differences.

NEW FORMULAS

$$SS_{total} = \Sigma X^2 - \frac{G^2}{N} \qquad df_{total} = N - 1$$

$$SS_{within} = \Sigma SS_{each\ treatment} \qquad df_{within} = \Sigma df_{each\ treatment} = N - k$$

$SS_{between}$ can be found by subtraction: $SS_{between} = SS_{total} - SS_{within}$ or computed:

$$SS_{between} = \Sigma \frac{T^2}{n} - \frac{G^2}{N} \quad \text{or} \quad n(SS_{means}) \qquad df_{between} = k - 1$$

$$MS = \frac{SS}{df} \qquad F = \frac{MS_{between}}{MS_{within}}$$

$$\eta^2 = \frac{SS_{between\ treatments}}{SS_{total}}$$

STEP BY STEP

<u>Analysis of Variance</u>: Analysis of variance is a hypothesis testing technique that is used to determine whether there are differences among the means of two or more populations. In this chapter we examined ANOVA for an independent-measures experiment, which means that the data consist of a separate sample for each treatment condition (or each population). Before you begin the actual analysis, you should complete all the preliminary calculations with the data, including T and SS for each sample, and G and ΣX^2 for the entire set of scores. The following example will be used to demonstrate ANOVA.

A researcher has obtained three different samples representing three populations. The data are as follows:

Sample 1	Sample 2	Sample 3	
0	6	6	
4	8	5	G = 60
0	5	9	
1	4	4	$\Sigma X^2 = 356$
0	2	6	
T = 5	T = 25	T = 30	
SS = 12	SS = 20	SS = 14	

Step 1: State the hypotheses and select an alpha level. The null hypothesis states that there are no mean differences among the three populations.

$H_0: \mu_1 = \mu_2 = \mu_3$

Remember, we generally do not try to list specific alternatives, but rather state a generic alternative hypothesis.

H_1: At least one population mean is different from another.

For this test we will use $\alpha = .05$.

Step 2: Locate the critical region. With k = 3 samples, the numerator of the F-ratio will have $df_{between}$ = k - 1 = 2. There are n = 5 scores in each sample. Within each sample there are n - 1 = 4 degrees of freedom, and summing across all three samples gives df_{within} = 4 + 4 + 4 = 12 for the denominator of the F-ratio. Thus, the F-ratio for this analysis will have df = 2, 12.

Sketch the entire distribution of F-ratios with df = 2, 12 and locate the extreme 5% of the distribution. The critical F value is 3.88.

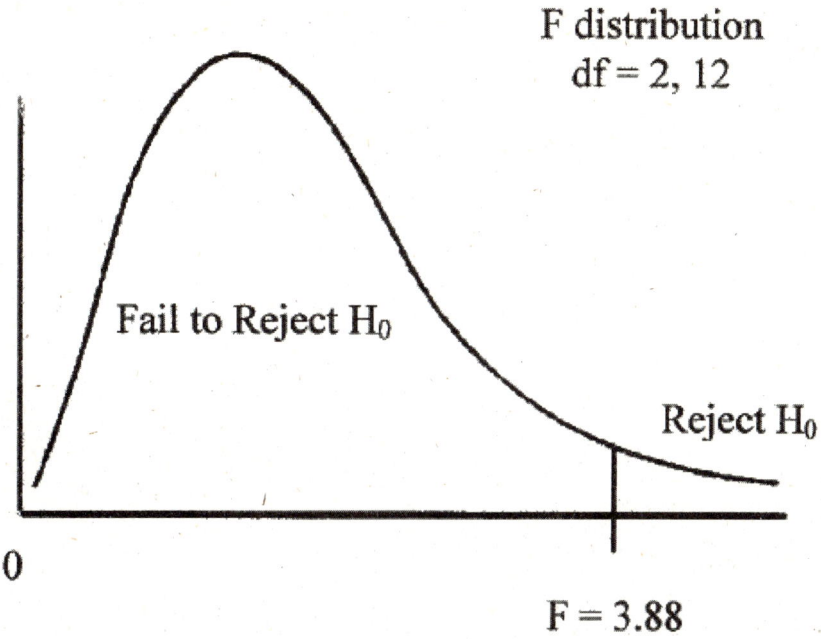

Step 3: Compute the test statistic. It is best to work through the calculations in a systematic way. Compute all three parts of the analysis (total, between, and within) and check that the two components add to the total.

$$SS_{total} = \Sigma X^2 - \frac{G^2}{N} = 356 - \frac{(60)^2}{15}$$

$$= 356 - 240$$

$$= 116$$

$$SS_{within} = \Sigma SS = 12 + 20 + 14 = 46$$

SS_{bet} can be found by subtraction: $SS_{total} - SS_{within} = 116 - 46 = 70$
or, by calculation

$$SS_{bet} = \Sigma \frac{T^2}{n} - \frac{G^2}{N} = \frac{(5)^2}{5} + \frac{(25)^2}{5} + \frac{(30)^2}{5} - \frac{(60)^2}{15}$$

$$= 5 + 125 + 180 - 240$$

$$= 70$$

INTRODUCTION TO ANALYSIS OF VARIANCE

We have already found df$_{between}$ and df$_{within}$. To complete the analysis of df, compute

$$df_{total} = N - 1 = 15 - 1 = 14$$
(Check that 14 = 12 + 2)

Next, compute the two variances (Mean Squares) that will form the F-ratio.

$$MS_{between} = \frac{SS_{between}}{df_{between}} = \frac{70}{2} = 35$$

$$MS_{within} = \frac{SS_{within}}{df_{within}} = \frac{46}{12} = 3.83$$

Finally, the F-ratio for these data is,

$$F = \frac{MS_{between}}{MS_{within}} = \frac{35}{3.83} = 9.14$$

Step 4: Make decision. The F-ratio for these data is in the critical region. The numerator is more than 9 times larger than the denominator which indicates a significant treatment effect. Reject H$_0$ and conclude that there are differences among the means of the three populations.

Measuring effect size with η^2

The data from the preceding example will be used to demonstrate the calculation of η^2 to measure effect size. For these data, SS$_{between\ treatments}$ = 70 and SS$_{total}$ = 116, so the percentage of variance accounted for by the treatment effects is

$$\eta^2 = \frac{SS_{between\ treatments}}{SS_{total}} = \frac{70}{116} = 0.603 \quad (60.3\%)$$

HINTS AND CAUTIONS

1. It may help you to understand analysis of variance if you remember that measuring variance is conceptually the same as measuring differences. The goal of the analysis is to determine whether the mean differences in the data are larger than would be expected if there was no treatment effect.
2. The formulas for SS between treatments and SS within treatments, and their role in the F-ratio, may be easier to remember if you look at the similarities between the independent-measures t formula and the F-ratio formula.
   a. The numerator of the t statistic measures the difference between the two sample means, $(M_1 - M_2)$. The numerator of the F-ratio also looks at

differences between treatments by computing the variability for the treatment totals.

b. The standard error in the denominator of the t statistic is computed by first pooling the two sample variances. This calculation uses the SS values from each of the two separate samples. The denominator of the F-ratio also uses the SS values from each of the separate samples to compute SS within treatments. In fact, when there are only two treatment conditions, the pooled variance from the t statistic is equivalent to the MS within treatments from the F-ratio.

SELF TEST

True/False Questions

1. The primary advantage of analysis of variance compared to t tests, is that the ANOVA can evaluate mean differences for studies that produce more than two means.

2. The larger the differences among the sample means, the larger the numerator of the F-ratio will be.

3. For an analysis of variance comparing three treatment means, H_0 states that all three population means are the same and H_1 states that all three population means are different.

4. For an analysis of variance comparing four treatment means with a separate sample of n = 8 participants in each treatment, the F-ratio will have df = 3, 31.

5. An F-ratio near 1.00 is an indication that there are no significant treatment effects.

6. An F-ratio can never be a negative number.

7. For an analysis of variance comparing three treatment means with a separate sample of n = 10 participants in each treatment, df_{total} = 29.

8. Effect size for analysis of variance is measured by η^2 which equals $SS_{between}$ divided by SS_{within}.

9. Post tests such as Scheffé are not necessary for an analysis of variance comparing only two treatment conditions.

10. In an analysis of variance, if all other factors are held constant, the larger the sample variances, the bigger the value for the F-ratio.

Multiple-Choice Questions

1. In an analysis of variance, which of the following is not true?
   a. $SS_{total} = SS_{between} + SS_{within}$
   b. $df_{total} = df_{between} + df_{within}$
   c. $MS_{total} = MS_{between} + MSS_{within}$
   d. All three choices (a, b, and c) are true.

2. When comparing more than two treatment means, why should you use an analysis of variance instead of using several t tests?
   a. Using several t tests increases the risk of a Type I error.
   b. Using several t tests increases the risk of a Type II error.
   c. The analysis of variance is more likely to detect a treatment effect.
   d. There is no advantage to using an analysis of variance instead of several t tests.

3. Which of the following is expected if the null hypothesis is true for an analysis of variance?
   a. $SS_{between}$ should be about the same size as SS_{total}
   b. $SS_{between}$ should be about the same size as SS_{within}
   c. $MS_{between}$ should be about the same size as MS_{total}
   d. $MS_{between}$ should be about the same size as MS_{within}

4. If an analysis of variance is used for the following data, what would be the effect of changing the value of M_1 to 20?   Sample Data
   a. increase $SS_{between}$ and increase the size of the F-ratio $M_1 = 15$ $M_2 = 25$
   b. increase $SS_{between}$ and decrease the size of the F-ratio. $SS_1 = 90$ $SS_2 = 70$
   c. decrease $SS_{between}$ and increase the size of the F-ratio
   d. decrease $SS_{between}$ and decrease the size of the F-ratio

5. If an analysis of variance is used for the following data, what would be the effect of changing the value of SS_2 to 100?   Sample Data
   a. increase SS_{within} and increase the size of the F-ratio $M_1 = 15$ $M_2 = 25$
   b. increase SS_{within} and decrease the size of the F-ratio $SS_1 = 90$ $SS_2 = 70$
   c. decrease SS_{within} and increase the size of the F-ratio
   d. decrease SS_{within} and decrease the size of the F-ratio

6. The following table shows the results of an analysis of variance comparing three treatment conditions with a sample of $n = 10$ participants in each treatment. Note that several values are missing in the table. What is the missing value for SS_{total}?

   a. 22
   b. 30
   c. 54
   d. 74

Source	SS	df	MS	
Between	20	xx	xx	F = xx
Within	xx	xx	2	
Total	xx	xx		

7. The following table shows the results of an analysis of variance comparing two treatment conditions with a sample of n = 11 participants in each treatment. Note that several values are missing in the table. What is the missing value for the F-ratio?

    a. 2
    b. 7
    c. 14
    d. 28

Source	SS	df	MS	
Between	xx	xx	14	F = xx
Within	xx	xx	xx	
Total	154	xx		

8. For an analysis of variance comparing four treatments, $MS_{between}$ = 12. What is the value of $SS_{between}$?

    a. 3
    b. 4
    c. 36
    d. 48

9. A researcher uses analysis of variance to test for mean differences among four treatments with a sample of n = 6 in each treatment. The F-ratio for this analysis would have what df values?

    a. df = 3, 5
    b. df = 3, 15
    c. df = 3, 20
    d. df = 4, 24

10. A researcher obtains an F-ratio with df = 2, 36 from an independent-measures research study. What is the total number of individuals who participated in the study?

    a. N = 3
    b. N = 37
    c. N = 39
    d. N = 40

11. In an analysis of variance, large mean differences from one sample to another will produce a large value for _____.

    a. $SS_{between\ treatments}$
    b. $SS_{within\ treatments}$
    c. SS_{total}
    d. Large mean differences will cause all three SS values to be large.

12. In analysis of variance, large values for the sample variances will produce a large value for_____.

    a. $SS_{between\ treatments}$
    b. $SS_{within\ treatments}$
    c. SS_{total}
    d. Large sample variances will cause all three SS values to be large.

13. In an ANOVA, which of the following is most likely to produce a large value for the F-ratio?
   a. large mean differences and small sample variances
   b. large mean differences and large sample variances
   c. small mean differences and small sample variances
   d. small mean differences and large sample variances

14. An independent-measures t test produced a t statistic with df = 20. If the same data had been evaluated with an analysis of variance, what would be the df values for the F-ratio?
   a. 1, 19
   b. 1, 20
   c. 2, 19
   d. 2, 20

15. The following table shows the results of an analysis of variance. Based on this table, what is the value for 0^2, the percentage of variance accounted for?
   a. 16/18
   b. 16/100
   c. 84/100
   d. 4/10

Source	SS	df	MS	
Between	16	2	8	F = 2.00
Within	84	21	4	
Total	100	23		

Other Questions

1. The following data summarize the results of two experiments. Each experiment compares three treatment conditions, and each experiment uses separate samples of n = 10 for each treatment.

Experiment A			Experiment B		
Treatment			Treatment		
1	2	3	1	2	3
M = 1	M = 3	M = 5	M = 1	M = 10	M = 20
s = 15	s = 12	s = 18	s = 3	s = 5	s = 4

Just looking at the data - without doing any calculations - answer each of the following questions.
   a. Which experiment will produce the larger $MS_{between}$?
   b. Which experiment will produce the larger MS_{within}?
   c. Which experiment will produce the larger F-ratio?

2. Use an analysis of variance with $\alpha = .05$ to determine whether the following data provide evidence of any significant differences among the three treatments.

Treatments			
I	II	III	
0	4	1	$G = 30$
2	6	0	
2	1	3	$\Sigma X^2 = 114$
0	5	1	
1	4	0	
T = 5	T = 20	T = 5	
SS = 4	SS = 14	SS = 6	

3. A researcher conducts an experiment comparing three treatment conditions with a separate sample of n = 8 in each treatment. An analysis of variance is used to evaluate the data, and the results of the ANOVA are presented in the table below. Complete all missing values in the table.

Source	SS	df	MS	
Between Treatments	12	__	__	F = 2.00
Within Treatments	__	__	__	
Total	__	__		

ANSWERS TO SELF TEST

True/False Answers

1. True
2. True
3. False. The alternative hypothesis says that at least one mean is different from another.
4. False. df = 3, 28
5. True
6. True
7. True
8. False. Effect size is $SS_{between}$ divided by SS_{total}.
9. True
10. False. Large variance contributes to the denominator and produces a smaller F-ratio.

Multiple-Choice Answers

1. c The SS and df are analyzed into two parts, not the MS values.
2. a Each t test has its own risk of a Type I error.
3. d The two variances in the F-ratio should be similar.

4. d   It will reduce the size of the mean difference in the numerator.
5. b   In will increase the within treatments variance in the denominator.
6. d $df_{within} = 27$, $SS_{within} = 54$, and $SS_{total} = 20 + 54$
7. a $df_{within} = 20$, $SS_{within} = 140$, and $F = 14/7$
8. c $MS = SS/df$
9. c   df between $= k - 1 = 3$ and df within $= N - k = 20$.
10. c   df total $= 2 + 36 = 38 = N - 1$.
11. a   The between treatments SS measures the size of the mean differences.
12. b   The within treatments SS adds the SS values from within the treatments.
13. a   Mean differences contributed to the numerator and variance contributes to the denominator of the F-ratio.
14. b   df between $= 2 - 1 = 1$, and df within $= (n_1 - 1) + (n_2 - 1)$
15. b   Eta squared equals SS between divided by SS total.

Other Answers

1. a. Experiment B has larger mean differences which will produce a larger $MS_{between}$.
   b. Experiment A has larger sample standard deviations (more variance within samples) and will produce a larger MS_{within}.
   c. Experiment B has larger differences between treatments and smaller variability within treatments. This combination will produce a larger F-ratio.

2. The ANOVA is summarized as follows:

Source	SS	df	MS	
Between Treatments	30	2	15	$F = 7.50$
Within Treatments	24	12	2	
Total	54	14		

With df = 2, 12 the critical value is F = 3.88. Reject H_0 and conclude that there are significant differences among the three treatments.

3.

Source	SS	df	MS	
Between Treatments	12	2	6	$F = 2.00$
Within Treatments	63	21	3	
Total	75	23		

CHAPTER 14

REPEATED-MEASURES ANALYSIS OF VARIANCE

CHAPTER SUMMARY

The goals of Chapter 14 are:
1. To introduce the logic underlying the repeated-measures analysis of variance.
2. To demonstrate the similarities and differences between a repeated-measures ANOVA and an independent-measures ANOVA.
3. To demonstrate the measurement of effect size for the repeated-measures ANOVA.

The Logical Background for a Repeated-Measures ANOVA

Chapter 14 extends analysis of variance to research situations using repeated-measures (or related-samples) research designs. Much of the logic and many of the formulas for repeated-measures ANOVA are identical to the independent-measures analysis introduced in Chapter 13. However, the repeated-measures ANOVA includes a second stage of analysis in which variability due to **individual differences** is subtracted out of the error term. The repeated-measures design eliminates individual differences from the between-treatments variability because the same subjects are used in every treatment condition. To balance the F-ratio, however, the calculations require that individual differences also be eliminated from the denominator of the F-ratio. The result is a test statistic similar to the independent-measures F-ratio but with all individual differences removed.

Comparing Independent-Measures and Repeated-Measures ANOVA

In Chapter 13, we introduced the independent-measures analysis of variance. The independent-measures analysis is used in research situations for which there is a separate sample for each treatment condition. The analysis compares the mean square (MS) between treatments to the mean square within treatments in the form of a ratio

$$F = \frac{MS_{\text{between treatments}}}{MS_{\text{within treatments}}}$$

Now, in Chapter 14, we are evaluating repeated-measures studies in which the same sample serves in all of the different treatment conditions. What makes the repeated-measures analysis different from the independent-measures analysis is the treatment of variability from individual differences. Recall that the independent-measures F ratio (Chapter 13) has the following structure:

$$F = \frac{MS_{\text{between treatments}}}{MS_{\text{within}}} = \frac{\text{treatment effect} + \text{error (including individual differences)}}{\text{error (including individual differences)}}$$

In this formula, when the treatment effect is zero (H_0 true), the expected F ratio is one.

In the repeated-measures study, there are no individual differences between treatments because the same individuals are tested in every treatment. This means that variability due to individual differences is not a component of the numerator of the F ratio. Therefore, the individual differences must also be removed from the denominator of the F ratio to maintain a balanced ratio with an expected value of 1.00 when there is no treatment effect. That is, we want the repeated-measures F-ratio to have the following structure:

$$F = \frac{\text{treatment effect} + \text{error (without individual differences)}}{\text{error (with individual differences removed)}}$$

This is accomplished by a two-stage analysis. In the first stage, total variability (SS_{total}) is partitioned into the between-treatments SS and within-treatments SS. (Note, this is the same process we used for the independent-measures ANOVA.) The components for between-treatments variability are the treatment effect (if any) and error. Individual differences do not appear here because the same sample of subjects serves in every treatment. On the other hand, individual differences do play a role in SS_{within} because the sample contains different subjects.

In the second stage of the analysis, we measure the individual differences by computing the variability between subjects, or $SS_{between\ subjects}$. This value is subtracted from SS_{within} leaving a remainder, variability due to experimental error, SS_{error}. This two stage process is outlined in Figure 14.2 of your text.

A similar two-stage process is used to analyze the degrees of freedom (Figure 14.2 of your text). For the repeated-measures analysis, the mean square values and the F-ratio are as follows:

$$MS_{between\ treatments} = \frac{SS_{between\ treatments}}{df_{between\ treatments}}$$

$$MS_{error} = \frac{SS_{error}}{df_{error}} \qquad F = \frac{MS_{between\ treatments}}{MS_{error}}$$

One of the main advantages of the repeated-measures design is that the role of individual differences can be eliminated from the study. This advantage can be very important in situations where large individual differences would otherwise obscure the treatment effect in an independent-measures study.

Measuring Effect Size for the Repeated-Measures Analysis of Variance

In addition to determining the significance of the sample mean differences with a hypothesis test, it is also recommended that you determine the size of the mean differences by computing a measure of effect size. As we noted in Chapter 13, the common technique for measuring effect size for an analysis of variance is to compute the percentage of variance that is accounted for by the treatment effects. In the context of

ANOVA this percentage is identified as η^2 (the Greek letter eta, squared). Before computing η^2, however, it is customary to remove any variability that is accounted for by factors other than the treatment effect. In the case of a repeated-measures design, part of the variability is accounted for by individual differences and can be measured with $SS_{between\ subjects}$. When the variability due to individual differences is subtracted out, the value for η^2 then determines how much of the remaining, unexplained variability is accounted for by the treatment effects. Because the individual differences are removed from the total SS before eta squared is computed, the resulting value is often called a *partial* eta squared. The formula for computing effect size for a repeated-measures ANOVA is:

$$\eta^2 = \frac{SS_{between\ treatments}}{SS_{total} - SS_{between\ subjects}} = \frac{SS_{between\ treatments}}{SS_{error} + SS_{between\ treatments}}$$

LEARNING OBJECTIVES

1. Be able to explain the logic of the repeated-measures ANOVA.

2. Understand how variability is partitioned and what sources of variability contribute to each component.

3. Know the differences between the analysis for repeated- versus independent-measures designs.

4. Be able to perform the computations for a complete analysis.

5. Be able to compute η^2 to measure effect size for the repeated-measures ANOVA.

NEW TERMS AND CONCEPTS

The following terms were introduced in this chapter. You should be able to define or describe each term and, where appropriate, describe how each term is related to other terms in the list.

Between-treatments variability	The differences that exist from one treatment to another (a measure of mean differences).
Within-treatments variability	The differences that exist inside each treatment condition.
Between-subjects variability	The differences that exist from one subject to another.
Error variability	Unexplained, unsystematic differences that are not caused by any known factor.

Treatment effect	Systematic differences that are caused by changing treatment conditions.

Individual differences	Consistent differences that exist between one individual and another.

F-ratio	A ratio of two variances: between-treatments variance in the numerator and chance/error in the denominator.

NEW FORMULAS

In addition to the total SS, between-treatments SS, within-treatments SS, and the corresponding df and MS values that were presented in Chapter 13, several new formulas are introduced in Chapter 14:

$$SS_{\text{between subjects}} = \Sigma \frac{P^2}{k} - \frac{G^2}{N}$$

$$df_{\text{between subjects}} = n - 1$$

$$SS_{\text{error}} = SS_{\text{within}} - SS_{\text{between subjects}}$$

$$df_{\text{error}} = df_{\text{within}} - df_{\text{between subjects}}$$
$$= (N - k) - (n - 1)$$

$$\eta^2 = \frac{SS_{\text{between treatments}}}{SS_{\text{total}} - SS_{\text{between subjects}}} = \frac{SS_{\text{between treatments}}}{SS_{\text{error}} + SS_{\text{between treatments}}}$$

STEP BY STEP

The repeated-measures ANOVA is used to determine whether there are any differences among the means of two or more different treatments, using the data from a single sample that has been measured in each treatment condition. The calculations and notation for the repeated-measures ANOVA are very similar to the independent-measures design. In fact, the first stage of the repeated-measures analysis is identical to the independent ANOVA. However, the repeated analysis continues through a second stage where the individual differences are removed from the denominator of the F-ratio. The following example will be used to demonstrate the repeated-measures ANOVA.

A researcher is comparing four different treatment conditions using a repeated-measures experiment with a sample of n = 6 subjects. The data from this experiment are as follows:

Treatment

Subject	1	2	3	4	P
#1	0	2	6	0	8
#2	4	0	7	1	12
#3	5	4	11	0	20
#4	8	7	13	4	32
#5	7	2	9	2	20
#6	6	3	14	5	28
T	30	18	60	12	
SS	40	28	52	22	

$G = 120 \quad \Sigma X^2 = 970$

Step 1: State the hypotheses and select an alpha level. The null hypothesis states that there are no differences among the means for the four treatment conditions.

H_0: $\mu_1 = \mu_2 = \mu_3 = \mu_4$

The general alternative hypothesis is,
  H_1: At least one of the treatment means is different from another.
We will use $\alpha = .05$.

Step 2: Locate the critical region. The first problem is to determine the df values for the F-ratio. We will conduct a complete analysis of df to determine the values for $df_{between\ treatments}$ and df_{error}. For these data,

$df_{total} = N - 1 = 24 - 1 = 23$
$df_{between\ treatments} = k - 1 = 4 - 1 = 3$
$df_{within\ treatments} = N - k = 24 - 4 = 20$

This completes the first stage of the analysis. (Check to be sure that the total equals the sum of the two components.) For the second stage,

$df_{between\ subjects} = n - 1 = 6 - 1 = 5$

$df_{error} = df_{within} - df_{between\ subjects}$
$\qquad = 20 - 5$
$\qquad = 15$

REPEATED-MEASURES ANALYSIS OF VARIANCE

The F-ratio will have df = 3, 15. Sketch the entire distribution of F-ratios with df = 3,15 and locate the extreme 5% of the distribution. The critical F value is 3.29.

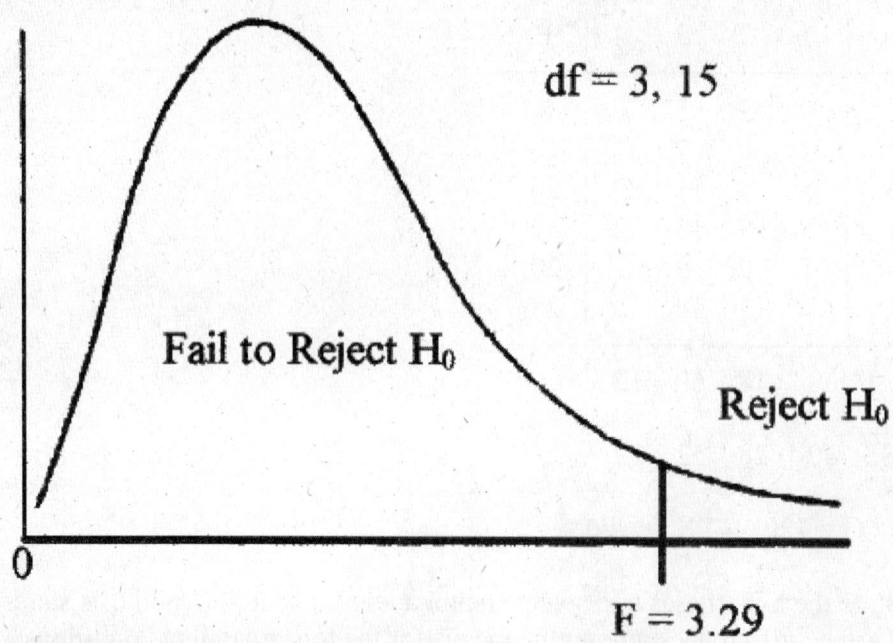

Step 3: Compute the test statistic. We have completed the analysis for df, so we will continue with the analysis of SS. As before, we will complete the analysis in two stages.

$$SS_{total} = \Sigma X^2 - \frac{G^2}{N}$$

$$= 970 - \frac{120^2}{24} = 970 - 600 = 370$$

$SS_{within} = \Sigma SS = 40 + 28 + 52 + 22 = 142$
Note: $SS_{between}$ can be found by subtraction,
$SS_{between} = SS_{total} - SS_{within}$
or can be computed directly as follows:

$$SS_{\text{between treatments}} = \Sigma \frac{T^2}{n} - \frac{G^2}{N}$$

$$= \frac{30^2}{6} + \frac{18^2}{6} + \frac{60^2}{6} + \frac{12^2}{6} - \frac{120^2}{24}$$

$$= 150 + 54 + 600 + 24 - 600$$

$$= 228$$

This completes the first stage. (Check to be sure that the two components add to the total.) Continuing with the second stage,

$$SS_{\text{subjects}} = \Sigma \frac{P^2}{k} - \frac{G^2}{N}$$

$$= \frac{8^2}{4} + \frac{12^2}{4} + \frac{20^2}{4} + \frac{32^2}{4} + \frac{20^2}{4} + \frac{28^2}{4} - \frac{120^2}{24}$$

$$= 16 + 36 + 100 + 256 + 100 + 196 - 600$$

$$= 704 - 600 = 104$$

$$SS_{\text{error}} = SS_{\text{within}} - SS_{\text{subjects}}$$

$$= 142 - 104$$

$$= 38$$

Next, compute the two mean squares (MS values) that will form the F-ratio.

$$MS_{\text{between treatments}} = \frac{SS_{\text{between treatments}}}{df_{\text{between treatments}}} = 228/3 = 76$$

$$MS_{\text{error}} = \frac{SS_{\text{error}}}{df_{\text{error}}} = 38/15 = 2.53$$

Finally, the F-ratio for these data is,

$$F = \frac{MS_{\text{between treatments}}}{MS_{\text{error}}} = 76/2.53 = 30.04$$

Step 4: Make decision. The F-ratio for these data is in the critical region. This is a very unlikely value to be obtained if H_0 is true. Therefore, we reject H_0 and conclude that there are significant differences among the four treatment means.

Measuring Effect Size for the Repeated-Measures ANOVA

The data from the preceding example will be used to demonstrate the calculation of η^2 to measure effect size. For these data, $SS_{between\ treatments} = 228$, $SS_{total} = 370$, and $SS_{between\ subjects} = 104$. Removing the variability that is explained by individual differences, we obtain

$$SS_{total} - SS_{between\ subjects} = 370 - 104 = 266 \text{ points}$$

This is the variability that is not explained by factors other than the treatment. Of this amount, the treatment accounts for $SS_{between\ treatments} = 228$ points. Expressed as a proportion,

$$\eta^2 = \frac{228}{266} = 0.857 \quad (\text{or } 85.7\%)$$

Using the alternative equation, we obtain exactly the same result:

$$\eta^2 = \frac{SS_{between\ treatments}}{SS_{error} + SS_{between\ treatments}} = \frac{228}{38 + 228} = \frac{228}{266} = 0.857$$

HINTS AND CAUTIONS

1. In the repeated-measures ANOVA, it is important that you remember to use the correct error term (denominator) for the F-ratio, namely MS_{error}.

2. A very common mistake when locating the critical region for a repeated-measures F-ratio is to use within-treatments df. Remember, the correct df value for the error term is the error df.

3. One important fact that may help you to remember the structure and the value of a repeated-measures ANOVA is that the repeated-measures design eliminates individual differences. The individual differences do not exist in the numerator of the F-ratio because the same individuals are used in all of the treatment conditions. In the denominator of the F-ratio there are no individual differences because they are subtracted out in the second stage of the analysis.

4. In a typical display of data for a repeated-measures experiment the scores for each subject are listed in rows and the scores for each treatment are listed in columns (see the Step-by-Step example). When you are computing $SS_{between}$ you are measuring the differences between columns of data. When you compute $SS_{subjects}$ you are measuring the differences between rows of data. This observation should help you recognize the similarity between the formulas for these two SS values.

$$SS_{between} = \Sigma \frac{T^2}{n} - \frac{G^2}{N} \qquad SS_{subjects} = \Sigma \frac{P^2}{k} - \frac{G^2}{N}$$

SELF TEST

True/False Questions

1. For a repeated-measures design, the differences between treatments may be caused by a treatment effect or by sampling error (random factors), but they cannot be caused by individual differences.

2. The denominator of the repeated-measures F-ratio is intended to measure differences that exist without any systematic treatment effect or any systematic individual differences.

3. The analysis of total variability into between-treatments and within-treatments variability is the same for a repeated-measures ANOVA and an independent-measures ANOVA.

4. For a repeated-measures study, if there are no systematic treatment effects (H_0 is true), then the numerator and denominator of the F-ratio are both measuring the same sources of variance.

5. In a repeated-measures ANOVA the variability caused by systematic individual differences must be measured and subtracted out of the between-treatments variability in the numerator of the F-ratio.

6. A repeated-measures study comparing three treatment conditions with a sample of $n = 10$ participants would produce an F-ratio with $df = 2, 18$.

7. A researcher reports the results of an ANOVA for a repeated-measures study as "$F(2, 12) = 7.84, p < .01$." This study used a sample of $n = 13$ participants.

8. A repeated-measures ANVOA produces $SS_{between\ treatments} = 30$ and $SS_{total} = 60$. For this study, the percentage of variance accounted for by the treatments is $\eta = 30/60 = 0.50$.

9. For the repeated-measures analysis, the F-ratio, on average, is expected to be zero when H_0 is true.

10. For the repeated ANOVA, $SS_{within\ treatments} = SS_{between\ subjects} + SS_{error}$.

Multiple-Choice Questions

1. For a repeated-measures ANOVA, why aren't individual differences a source of variability for the numerator of the F-ratio.
   a. They are measured and subtracted out.
   b. The same individuals participate in all of the treatment conditions.
   c. The individual differences in the numerator are canceled out by the individual differences in the denominator.
   d. Individual difference are a source of variability in the numerator.

2. The following table shows the results of a repeated-measures analysis of variance comparing three treatment conditions with a sample of n = 10 participants. Note that several values are missing in the table. What is the missing value for the F-ratio?

   a. 4
   b. 7
   c. 10
   d. 14

Source	SS	df	MS	
Between	28	xx	xx	F = xx
Within	xx	xx		
Bet. Sub.	16	xx		
Error	xx	xx	xx	
Total	80	xx		

3. The following table shows the results of a repeated-measures analysis of variance comparing two treatment conditions with a sample of n = 12 participants. Note that several values are missing in the table. What is the missing value for SS_{total}?

   a. 45
   b. 47
   c. 80
   d. 148

Source	SS	df	MS	
Between	xx	xx	12	F = 4.00
Within	xx	xx		
Bet. Sub.	35	xx		
Error	xx	xx	xx	
Total	xx	xx		

4. The following table shows the results of a repeated-measures ANOVA. Based on this table, what is the value for 0^2, the percentage of variance accounted for by the treatments?

   a. 22/30
   b. 22/50
   c. 22/58
   d. 22/80

Source	SS	df	MS	
Between	22	2	11	F = 5.50
Within	58	21		
Bet. Sub.	30	7		
Error	28	14	2	
Total	80	23		

5. In the repeated-measures ANOVA, individual differences are *not* a source of variability for the error SS because _____.
   a. this factor is mathematically subtracted out
   b. the same subjects serve in every treatment condition
   c. individual differences in the numerator and denominator of the F ratio cancel each other
   d. None of the other 3 choices is correct.

6. Which of the following accurately describes the two stages of a repeated-measures analysis of variance?
    a. The first stage is identical to the independent-measures analysis and the second stage removes individual differences from the numerator of the F-ratio.
    b. The first stage is identical to the independent-measures analysis and the second stage removes individual differences from the denominator of the F-ratio.
    c. The first stage removes individual differences from the numerator of the F-ratio
    and the second stage is identical to the independent-measures analysis.
    d. The first stage removes individual differences from the denominator of the F-ratio and the second stage is identical to the independent-measures analysis.

7. A repeated-measures ANOVA with n = 5 subjects has df within-treatment equal to 12. What is the value for df_{error} for this analysis?
    a. 8
    b. 16
    c. 48
    d. insufficient information to find error SS

Questions 8-10 refer to the following data from a repeated-measures study.

Subject	Treatment 1	Treatment 2	Treatment 3	P
A	0	2	1	3
B	1	1	4	6
C	2	3	4	9
D	1	2	3	6
	T = 4	T = 8	T = 12	
	SS = 2	SS = 2	SS = 6	

8. What are the degrees of freedom for the repeated-measures F-ratio?
    a. 2, 6
    b. 2, 9
    c. 2, 11
    d. 3, 11

9. What is the value for $SS_{between\ subjects}$?
    a. 2
    b. 3
    c. 6
    d. 10

10. What is the value for SS_{error}?
    a. 2
    b. 4
    c. 6
    d. 10

11. A researcher reports an F-ratio with df = 2, 40 from a repeated-measures ANOVA. How many treatment conditions were compared in this experiment?
    a. 2
    b. 3
    c. 4
    d. 41

12. A researcher reports an F-ratio with df = 2, 40 from a repeated-measures ANOVA. How many subjects participated in this experiment?
    a. 44
    b. 41
    c. 63
    d. 21

13. In a repeated-measures ANOVA, which of the following is not computed directly but rather is obtained by a process of subtraction?
    a. $SS_{Between\ Subjects}$
    b. $SS_{Within\ Treatments}$
    c. $SS_{Between\ Treatments}$
    d. SS_{Error}

14. A repeated-measures study uses a sample of n = 10 participants to evaluate the mean differences among four treatment conditions. In the analysis of variance for this study, what is the value for $df_{between\ subjects}$?
    a. 9
    b. 27
    c. 36
    d. 39

15. A repeated-measures analysis of variance for a study comparing three treatment conditions with a sample of n = 10 participants, produces an F-ratio of F = 5.40. For this result, which of the following is the correct statistical decision?
    a. Reject the null hypothesis with $\forall = .05$ but not with $\forall = .01$.
    b. Reject the null hypothesis with either $\forall = .05$ or $\forall = .01$.
    c. Fail to reject the null hypothesis with either $\forall = .05$ or $\forall = .01$.
    d. There is not enough information to determine the correct decision.

Other Questions

1. The final F-ratio in the repeated-measures analysis is structured so that it has an expected value of 1.00 when the null hypothesis is true.
    a. What sources of variability contribute to the numerator of the F-ratio ($MS_{between\ treatments}$)?
    b. Explain why "individual differences" is not listed as a source of variability in the numerator.
    c. What sources of variability contribute to the denominator of the F-ratio (MS_{error})?
    d. Explain why "individual differences" is not listed as a source of variability in the denominator.

2. An industrial researcher tests 3 different keyboard designs for a new computer to determine which one produces optimal performance. Four computer operators are given text material and are told to type the material as fast as they can. They spend 3 minutes on each keyboard with a 5 minute rest between each trial. The number of errors committed are recorded. Do the following data indicate a significant difference among the 3 keyboard types? Test at the .05 level of significance.

Operator	I	II	III	P
#1	6	2	4	12
#2	8	6	7	21
#3	3	6	9	18
#4	3	2	4	9
T	20	16	24	
SS	18	16	18	

$\Sigma X^2 = 360$

$G = 60$

3. A sample of n = 7 individuals is selected to participate in a learning study. Each individual is tested at five different stages during learning (after 1 hour, 2 hours, 3 hours, etc.). The data from this experiment were examined using a repeated-measures ANOVA to determine whether there was any evidence of a practice effect. The results are presented in the following summary table. Complete all the missing values in the table. Hint: Begin with the df column.

Source	SS	df	MS	
Between Treatments	___	___	10	F = ___
Within Treatments	___	___		
Between Subjects	52	___		
Error	___	___	___	
Total	140	___		

4. A toy manufacturer is testing 3 versions of a toy that is under development. Among other things, the manufacturer would like to see which version of the toy attracts the most attention. A psychologist allows a child to play with all 3 toys and records the amount of time (in minutes) spent playing with each. A sample of n = 5 children is used. Use the following data to determine whether there is a significant preference among the three toys. Use the .05 level of significance.

	Toy		
Child	1	2	3
A	0	2	4
B	2	2	8
C	3	1	5
D	0	3	6
E	0	2	7

ANSWERS TO SELF TEST

True/False Answers

1. True.
2. True
3. True
4. True
5. False. Individual differences do not exist between treatments because all treatments use exactly the same subjects.
6. True
7. False. The study used n = 7 subjects.
8. False. Variability from individual differences must be removed before the percentage is computed.
9. False. If the null hypothesis is true, the F-ratio should be near 1.00.
10. True

Multiple-Choice Answers

1. b Individual differences do not exist between treatments because all treatments use exactly the same subjects.
2. b $SS_{within} = 52$, $SS_{between\ subjects} = 36$, and $F = 14/2 = 7$
3. c $MS_{error} = 3$, $SS_{error} = 33$, $SS_{within} = 68$
4. b Eta squared is $SS_{treatment}$ divided by the sum of $MS_{treatnebt} + SS_{error}$
5. a. The individual differences are subtracted out in the second stage of the analysis.
6. b Because the same individuals are in every treatment, there are no individual differences between treatments, however, the individual differences must be subtracted from the denominator of the F-ratio.
7. a Error = Within treatment − between subjects.

8. a df = k − 1 and [(N − k) − (k − 1)] .
9. c Compute SS using the P values.
10. b $SS_{error} = SS_{within} − SS_{between\ subjects}$
11. b $df_{between} = 2 = k − 1$
12. d $SS_{error} = (k − 1)(n − 1)$
13. d SS error is the remainder when individual differences are subtracted from SS within treatments.
14. a between-subjects df = n − 1.
15. c With df = 2, 18 the critical value for .05 is 3.55 and for .01 it is 6.01.

Other Answers

1. a. The numerator of the F-ratio contains variability from treatment effects and random, unsystematic error.
   b. There are no individual differences in the variability between treatments because the same individuals are used in every treatment.
   c. The denominator of the F-ratio contains variability from random, unsystematic error.
   d. There are no individual differences in the denominator because they are subtracted out in the second stage of the analysis.

2. The null hypothesis states that there are no differences among the three keyboard types. With df = 2,6 the critical value is F = 5.14.

Source	SS	df	MS	
Between Treatments	8	2	4	F = 1.09
Within Treatments	52	9		
Between Subjects	30	3		
Error	22	6	3.67	
Total	60	11		

Fail to reject H_0. There are no significant differences among the three keyboard types.

3.

Source	SS	df	MS	
Between Treatments	40	4	10	F = 5.00
Within Treatments	100	30		
Between Subjects	52	6		
Error	48	24	2	
Total	140	34		

4. The null hypothesis states that there are no differences among the 3 toys. With df = 2,8, the critical boundary is F = 4.46.

Source	SS	df	MS	
Between Treatments	70	2	35	F = 20
Within Treatments	20	12		
Between Subjects	6	4		
Error	14	8	1.75	
Total	90	14		

Reject the null hypothesis and conclude that there are significant differences among the three toys.

CHAPTER 15

TWO-FACTOR ANALYSIS OF VARIANCE

CHAPTER SUMMARY

The goals of Chapter 15 are:
1. To introduce the concept of a two-factor research design.
2. To introduce the concepts of main effects and interactions.
3. To demonstrate the process of hypothesis testing with the two-factor ANOVA.
4. To demonstrate the calculation of effect size for main effects and interactions.

Two-Factor Designs

Chapter 15 extends analysis of variance to research designs that involve two independent variables. In the context of ANOVA, an independent variable (or a quasi-independent variable) is called a **factor**, and research studies with two factors are called **factorial designs** or simply **two-factor designs**. The two factors are identified as A and B, and the structure of a two-factor design can be represented as a matrix with the levels of factor A determining the rows and the levels of factor B determining the columns. For example, a researcher studying the effects of heat and humidity on performance could use the following experimental design:

	B1 80-degree room	B2 90-degree room	B3 100 degree room
A1 Low Humidity	Sample 1	Sample 2	Sample 3
A2 High Humidity	Sample 4	Sample 5	Sample 6

Notice that the study involves two levels of humidity and three levels of heat, creating a two-by-three matrix with a total of 6 different treatment conditions. Each treatment condition is represented by a cell in the matrix. For an independent-measures research study, a separate sample would be used for each of the six conditions.

Main Effects and Interactions

The goal for the two-factor ANOVA is to determine whether the mean differences that are observed for the sample data are sufficiently large to conclude that they are *significant* differences and not simply the result of sampling error. For the example we are considering, the goal is to determine whether different levels of heat and humidity produce significant differences in performance. To evaluate the sample mean differences, a two-factor ANOVA conducts three separate and independent hypothesis tests. The three tests evaluate:

1. **The Main Effect for Factor A** The mean differences between the levels of factor A are obtained by computing the overall mean for each row in the matrix. In this example, the main effect of factor A would compare the overall mean performance with high humidity versus the overall mean performance with low humidity.

2. **The Main Effect for Factor B** The mean differences between the levels of factor B are obtained by computing the overall mean for each column in the matrix. In this example, the ANOVA would compare the overall mean performance at 80E versus 90E versus 100E.

3. **The A x B Interaction** Often two factors will "interact" so that specific combinations of the two factors produce results (mean differences) that are not explained by the overall effects of either factor. For example, changes in humidity (factor A) may have a relatively small overall effect on performance when the temperature is low. However, when the temperature is high (100E), the effects of humidity may be exaggerated. In this case, unique combinations of heat and humidity produce results that are not explained by the overall main effects. These "extra" mean differences are the interaction.

The primary advantage of combining two factors (independent variables) in a single research study is that it allows you to examine how the two factors interact with each other. That is, the results will not only show the overall main effects of each factor, but also how unique combinations of the two variables may produce unique results. The interaction can be defined as "extra" mean differences, beyond the main effects of the two factors. An alternative definition is that an interaction exists when the effects of one factor depend on the levels of the second factor.

The Two-Factor Analysis of Variance

Each of the three hypothesis tests in a two-factor ANOVA will have its own F-ratio and each F-ratio has the same basic structure

$$F = \frac{\text{variance (differences) between means}}{\text{variance (differences) from error}} = \frac{MS_{\text{between treatments}}}{MS_{\text{within treatments}}}$$

Each MS value equals SS/df, and the individual SS and df values are computed in a two-stage analysis. The first stage of the analysis is identical to the single-factor ANOVA (Chapter 13) and separates the total variability (SS and df) into two basic components: between treatments and within treatments. The between-treatments variability measures the magnitude of the mean differences between treatment conditions (the individual cells in the data matrix) and is computed using the basic formulas for SS_{between} and df_{between}.

$$SS_{between} = \Sigma \frac{T^2}{n} - \frac{G^2}{N}$$

where the T values (totals) are the cell totals and n is the number of scores in each cell

$df_{between}$ = the number of cells (totals) minus one

The within-treatments variability measures the magnitude of the differences within each treatment condition (cell) and provides a measure of error variance, that is, unexplained, unpredicted differences due to error.

$$MS_{within} = \frac{SS_{within}}{df_{within}} = \frac{\Sigma SS}{\Sigma df} = \frac{SS_1 + SS_2 + SS_3 + ...}{df_1 + df_2 + df_3 + ...}$$

All three F-ratios use the same denominator, MS_{within}

The second stage of the analysis separates the between-treatments variability into the three components that will form the numerators for the three F-ratios: Variance due to factor A, variance due to factor B, and variance due to the interaction. Each of the three variances (MS) measures the differences for a specific set of sample means. The main effect for factor A, for example, will measure the mean differences between rows of the data matrix. The actual formulas for each SS and df are based on the sample totals (rather than the means) and all have the same structure:

$$SS_{between} = \Sigma \frac{T^2}{n} - \frac{G^2}{N}$$

$$= \Sigma \frac{(Total)^2}{number} - \frac{G^2}{N}$$

where the "number" is the number of scores that are summed to obtain each total

$df_{between}$ = the number of means (or totals) minus one

For factor A, the totals are the row totals and the df equals the number of rows minus 1.

For factor B, the totals are the column totals and the df equals the number of columns minus 1.

The interaction measures the "extra" mean differences that exist after the main effects for factor A and factor B have been considered. The SS and df values for the interaction are found by subtraction.

$$SS_{AxB} = SS_{bet\ cells} - SS_A - SS_B$$

$$df_{AxB} = df_{bet\ cells} - df_A - df_B$$

LEARNING OBJECTIVES

1. Be able to conduct a two-factor ANOVA to evaluate the data from an independent-measures experiment that uses two independent variables.

2. Understand the definitions of a main effect for one factor and an interaction between two factors, and be able to recognize an interaction from a description or a graph of experimental results.

3. Be able to perform a two-factor analysis of variance.

4. Be able to compute η^2 to measure effect size for main effects and for interactions.

NEW TERMS AND CONCEPTS

The following terms were introduced in this chapter. You should be able to define or describe each term and, where appropriate, describe how each term is related to other terms in the list.

Two-factor study	A research study examining two factors (two independent or quasi-independent variables).
Matrix and cells	A two-dimensional table is a matrix and each box in the table is called a cell.
Main effect	The overall mean differences between the levels of one factor. When the data are organized in a matrix, the main effects are the mean differences among the rows (or among the columns).
Interaction	Mean differences that cannot be explained by the main effects of the two factors. An interaction exists when the effects of one factor depend on the levels of the second factor.

NEW FORMULAS

$$SS_{total} = \Sigma X^2 - \frac{G^2}{N} \qquad df_{total} = N - 1$$

$$SS_{bet.\ cells} = \Sigma \frac{T^2}{n} - \frac{G^2}{N} \qquad df_{bet.\ cells} = (\text{number of cells} - 1)$$

$$SS_{within} = \Sigma SS_{each\ cell} \qquad df_{within} = \Sigma df_{each\ cell}$$

$$SS_{factor\ A} = \Sigma \frac{T^2_{ROW}}{n_{ROW}} - \frac{G^2}{N} \qquad df_{factor\ A} = (\text{number levels of A}) - 1$$

$$SS_{factor\ B} = \Sigma \frac{T^2_{COL}}{n_{COL}} - \frac{G^2}{N} \qquad df_{factor\ B} = (\text{number levels of B}) - 1$$

$$SS_{AxB} = SS_{bet\ cells} - SS_A - SS_B$$

$$df_{AxB} = df_{bet.\ cells} - df_A - df_B$$

To measure effect size for any treatment effect (either A or B or AxB)

$$\eta^2 = \frac{SS_{treatment\ effect}}{SS_{treatment\ effect} + SS_{within\ treatments}}$$

STEP BY STEP

<u>Two-Factor ANOVA</u>: The two-factor analysis of variance is used to evaluate mean differences in a research study that uses two independent variables or two quasi-independent variables. The two factors are generally identified as A and B, and the data are presented in a matrix with the levels of factor A determining the rows and the levels of factor B determining the columns. The analysis of variance evaluates three separate hypotheses: one concerning the main effect of factor A, one concerning the main effect of factor B, and one concerning the interaction. In this chapter we considered the two-factor analysis for an independent-measures research study which means that the

data consist of a separate sample for each AB treatment combination. The following example will be used to demonstrate the two-factor ANOVA.

The following data are from a two-factor experiment with 2 levels of factor A and 3 levels of factor B. There are n = 10 subjects in each treatment condition.

	B1	B2	B3	
A1	T = 10 SS = 20	T = 20 SS = 32	T = 30 SS = 35	$T_{A1} = 60$
A2	T = 10 SS = 15	T = 10 SS = 35	T = 10 SS = 25	$T_{A2} = 30$

$T_{B1} = 20 \quad T_{B2} = 30 \quad T_{B3} = 40$

$G = 90$
$\Sigma X^2 = 332$

Step 1: State the hypotheses and select an alpha level. Because the two-factor ANOVA evaluates three separate hypotheses, there will be three null hypotheses.

For Factor A: H_0: $\mu_{A1} = \mu_{A2}$ (no A-effect)
$\quad\quad\quad\quad\quad\quad H_1$: $\mu_{A1} \neq \mu_{A2}$

For Factor B: H_0: $\mu_{B1} = \mu_{B2} = \mu_{B3}$ (no B-effect)
$\quad\quad\quad\quad\quad\quad H_1$: At least one of the B means is different from another.

For AxB: H_0: There is no interaction between factors A and B. That is, the effect of either factor does not depend on the levels of the other factor.
$\quad\quad\quad\quad H_1$: There is an A x B interaction.

We will use $\alpha = .05$ for all three tests.

Step 2: Locate the critical regions. Because there are three separate tests, each with its own F-ratio, we will need to determine the critical region for each test separately. We begin by analyzing the degrees of freedom for these data to determine df for each F. The analysis proceeds in two stages.

df_{total} = N - 1 = 60 - 1 = 59
$df_{bet\ cells}$ = (number of cells) - 1 = 6 - 1 = 5
df_{within} = $\Sigma(n - 1)$ = 9 + 9 + 9 + 9 + 9 + 9 = 54

This completes the first stage of the analysis. (Check to be certain that the two components add to the total.)

Continuing with the second stage,

df_A = (number of levels of Factor A) − 1 = 2 − 1 = 1
df_B = (number of levels of Factor B) − 1 = 3 − 1 = 2
df_{AxB} = $df_{between\ cells}$ − df_A − df_B) = 5 − 1 − 2 = 2

Again, check to be certain that the three components from the second stage add to $df_{bet\ cells}$.

For these data, factor A will have an F-ratio with df = 1, 54. Factor B and the AxB interaction both will have F- ratios with df = 2, 54. Thus, we need F distributions and critical regions for two separate df values. Sketch the two distributions and locate the extreme 5% in each. (Because df = 54 is not listed, we have used 55 for the denominator in each case.)

F distribution for testing the main effect of Factor A

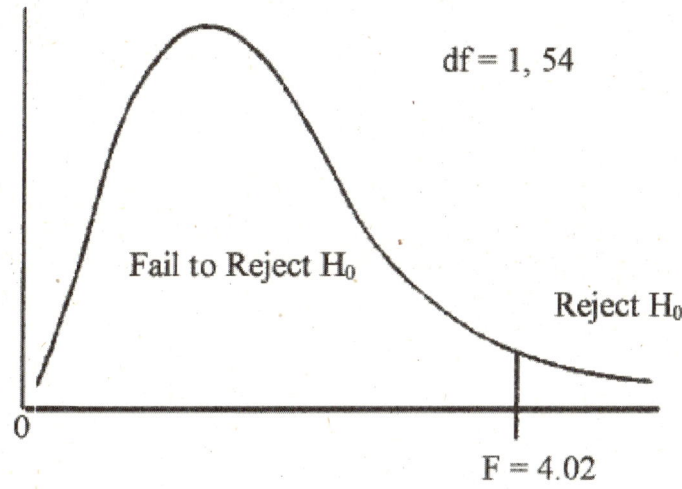

F distribution for testing the main effect of Factor B and the AxB interaction.

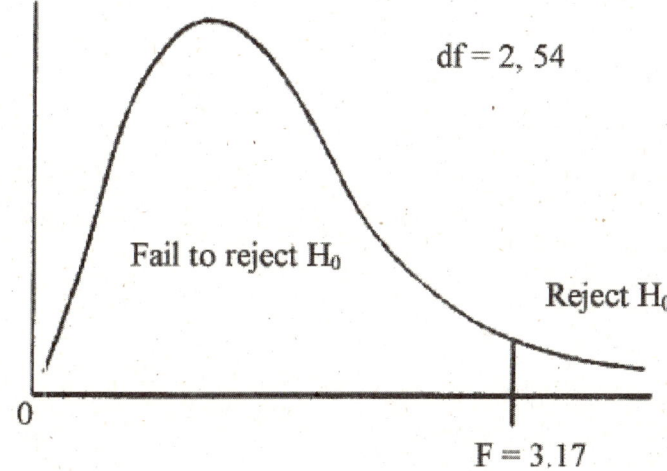

Step 3: Calculate the test statistic. Again, we will need three separate F-ratios. We have already analyzed the degrees of freedom for these data, so we will continue with the analysis of SS. As before, the analysis proceeds in two stages.

$$SS_{total} = \Sigma X^2 - \frac{G^2}{N}$$

$$= 332 - \frac{90^2}{60} = 332 - 135 = 197$$

$$SS_{bet.\ cells} = \Sigma \frac{T^2}{n} - \frac{G^2}{N}$$

$$= \frac{10^2}{10} + \frac{20^2}{10} + \frac{30^2}{10} + \frac{10^2}{10} + \frac{10^2}{10} + \frac{10^2}{10} - \frac{90^2}{60}$$

$$= 10 + 40 + 90 + 10 + 10 + 10 - 135$$

$$= 35$$

$$SS_{within} = \Sigma SS = 20 + 32 + 35 + 15 + 35 + 25$$

$$= 162$$

This completes stage one. Be sure that the two components add to the total.

$$SS_{total} = SS_{bet.\ cells} + SS_{within}$$
$$197 = 35 + 162$$

For the second stage,

$$SS_{factor\ A} = \Sigma \frac{T^2_{ROW}}{n_{ROW}} - \frac{G^2}{N}$$

$$= \frac{60^2}{30} + \frac{30^2}{30} - \frac{90^2}{60}$$

$$= 120 + 30 - 135$$

$$= 15$$

$$SS_{factor\ B} = \Sigma \frac{T^2_{COL}}{n_{COL}} - \frac{G^2}{N}$$

$$= \frac{20^2}{20} + \frac{30^2}{20} + \frac{40^2}{20} - \frac{90^2}{60}$$

$$= 20 + 45 + 80 - 135$$

$$= 10$$

$$SS_{AxB} = SS_{bet\ cells} - SS_A - SS_B$$

$$= 35 - 15 - 10$$

$$= 10$$

Again, check that these three components from stage two add to $SS_{bet\ cells}$.

Next, compute the MS values that will become the numerators for the three F-ratios.

$$MS_A = \frac{SS_A}{df_A} = 15/1 = 15$$

$$MS_B = \frac{SS_B}{df_B} = 10/2 = 5$$

$$MS_{AxB} = \frac{SS_{AxB}}{df_{AxB}} = 10/2 = 5$$

All three F-ratios will have the same error term denominator:

$$SS_{within}$$

TWO-FACTOR ANALYSIS OF VARIANCE

$$MS_{within} = \frac{}{df_{within}} = 162/54 = 3$$

Finally, the three F-ratios are,

$$\text{For Factor A: } F = \frac{MS_A}{MS_{within}} = 15/3 = 5.00$$

$$\text{For Factor B: } F = \frac{MS_B}{MS_{within}} = 5/3 = 1.67$$

$$\text{For AxB: } F = \frac{MS_{AxB}}{MS_{within}} = 5/3 = 1.67$$

Step 4: Make decision. The F-ratio for Factor A is in the critical region. Therefore, we reject this H_0 and conclude that there is a significant difference between the mean for A1 and the mean for A2. The F-ratios for factor B and for the AxB interaction are not in the critical region. Therefore, we conclude that there is no significant main effect for factor B, and the data are not sufficient to conclude that there is an interaction between factors A and B.

Measuring Effect Size: We will calculate η^2 (the percentage of variance accounted for) for factor A to demonstrate the measurement of effect size for a two-factor ANOVA. (Similar calculations could be used to measure effect size for factor B and for the AxB interaction.) For the data in the preceding example, $SS_A = 15$ and $SS_{within} = 162$.

$$\text{For factor A, } \eta^2 = \frac{SS_A}{SS_A + SS_{within}} = \frac{15}{15 + 162} = \frac{15}{177} = 0.085 \ (8.5\%)$$

Notice that the variability (SS) from factor B and from the AxB interaction are not included when computing the percentage of variability accounted for by factor A.

HINTS AND CAUTIONS

1. You should note that several of the SS formulas in the two-factor ANOVA have the same basic structure. Recognizing this structure can make it much easier to learn the formulas. For example, three of the SS formulas are computing variability due to differences between "things." These "things" and the corresponding SS values are:
   SS_A (between levels of factor A)
   SS_B (between levels of factor B)
   $SS_{bet\ cells}$ (between treatment conditions or cells)

The first term of each SS formula involves squaring a total and dividing by the number of scores that were added to compute the total. For example, SS_A squares each of the A totals and divides by n_A which is the number of scores used to find each A total. The second term in each of these SS formulas is G^2/N. Thus, all three of these formulas have the same structure that was used to compute SS between treatments for the single-factor ANOVA:

$$SS_{between} = \Sigma \frac{T^2}{n} - \frac{G^2}{N}$$

Note: You also could consider SS_{total} as measuring differences between scores. In this case each score is its own total, and $n = 1$, so the formula for SS_{total} also fits this same general structure. Also note that the degrees of freedom associated with each of these SS values can be determined by simply counting the number of "things" (or totals) and subtracting 1.

2. Remember that the F-ratios for factor A, factor B, and the AxB interaction can all have different values for df and therefore may have different critical values. Be sure that you use the appropriate critical region for each individual F-ratio.

SELF TEST

True/False Questions

1. A two-factor analysis of variance involves three separate F-ratios.

2. A two-factor study compares three different treatment conditions (factor 1) for males and females (factor 2). In this study, the main effect for gender is determined by the overall mean score for the males (averaged over the three treatments) and the corresponding overall mean score for the females.

3. A two-factor study compares two different treatment conditions (factor 1) for males and females (factor 2). In this study, the males have an average score of 20 in the first treatment and an average of 25 in the second. The females average 35 in the first treatment and 45 in the second. For this study, there is no interaction.

4. A doctor suspects that the effectiveness of a new cholesterol medication depends on the age of the patient. If the doctor is correct, the results from a two-factor study comparing medication versus no-medication for young versus old patients should produce an interaction.

5. A two-factor study compares 2 levels of factor A and 3 levels of factor B with a sample of $n = 5$ participants in each treatment condition. This study will use a total of 25 participants.

6. A two-factor study compares 2 levels of factor A and 3 levels of factor B with a sample of n = 10 participants in each treatment condition. For this study, the F-ratio for factor A has df = 2, 54.

7. A two-factor study compares 2 levels of factor A and 2 levels of factor B with a sample of n = 20 participants in each treatment condition. If the results are evaluated with a two-factor ANOVA, all the F-ratios will have df = 1, 76.

8. A two-factor analysis of variance with 2 levels of factor A and 3 levels of factor B involves six separate hypothesis tests.

9. If the A x B interaction is significant, then at least one of the two main effects also must be significant.

10. In a two-factor ANOVA for an independent-measures study, all of the F-ratios use the same denominator.

Multiple-Choice Questions

1. A two-factor study with two levels of factor A and three levels of factor B uses a separate group of n = 5 participants in each treatment condition. How many participants are needed for the entire study?
    a. 5
    b. 10
    c. 25
    d. 30

2. How many separate groups of participants would be needed for an independent-measures, two-factor study with 3 levels of factor A and 4 levels of factor B.
    a. 3
    b. 4
    c. 7
    d. 12

3. For a research study with 2 levels of factor A, 3 levels of factor B, and n = 5 in each treatment condition, what are the df values for the F-ratio evaluating the main effect for factor A?
    a. 1, 4
    b. 1, 24
    c. 1, 29
    d. 2, 29

4. In a two-factor ANOVA, what is the implication of a significant AxB interaction?
    a. At least one of the main effects must also be significant.
    b. Both of the main effects must also be significant.
    c. Neither of the two main effects can be significant.
    d. The significance of the interaction has no implications for the main effects.

5. The following table shows the results of a two-factor analysis of variance with two levels of factor A, three levels of factor B, and a separate sample of n = 5 participants in each of the treatment conditions. Note that several values are missing in the table. What is the missing value for the F-ratio for the AxB interaction?
    a. 2
    b. 4
    c. 8
    d. 16

Source	SS	df	MS	
Between	80	xx		
A	8	xx	xx	F = xx
B	xx	xx	20	F = xx
AxB	xx	xx	xx	F = xx
Within	xx	xx	xx	
Total	176	xx		

6. The following table shows the results of a repeated-measures ANOVA. Based on this table, what is the value for 0^2 for factor A?
    a. 12/36
    b. 12/96
    c. 12/99
    d. 12/120

Source	SS	df	MS	
Between	36	3		
A	12	1	12	F = 4.00
B	3	1	3	F = 1.00
AxB	21	1	21	F = 7.00
Within	84	28	3	
Total	120	31		

7. In a two-factor ANOVA, what is the implication of a significant AxB interaction?
    a. At least one of the main effects must also be significant.
    b. Both of the main effects must also be significant.
    c. Neither of the two main effects can be significant.
    d. The significance of the interaction has no implications for the main effects.

8. Which of the following is typically not calculated in a two-factor ANVOA?
    a. $MS_{between\ treatments}$
    b. $MS_{within\ treatments}$
    c. $MS_{for\ factor\ A}$
    d. $MS_{for\ the\ AxB\ interaction}$

9. The following data represent the means for each treatment condition in a two factor experiment. Note that one mean is not given. What value for the missing mean would result in no main effect for factor B?
   a. 20
   b. 30
   c. 40
   d. 50

	B1	B2
A1	20	10
A2	40	?

10. The following data represent the means for each treatment condition in a two factor experiment. Note that one mean is not given. What value for the missing mean would result in no AxB interaction?
   a. 10
   b. 20
   c. 30
   d. 40

	B1	B2
A1	20	30
A2	10	?

Questions 11 – 13 refer to the following data from a two-factor study.

	B1	B2
A1	$n = 10$, $T = 40$, $SS = 70$	$n = 10$, $T = 10$, $SS = 80$
A2	$n = 10$, $T = 30$, $SS = 73$	$n = 10$, $T = 20$, $SS = 65$

$N = 40$, $G = 100$, $G^2/N = 250$, $\Sigma X^2 = 588$

11. For these data, what is $df_{within\ treatments}$?
   a. 9
   b. 18
   c. 36
   d. 39

12. For these data, what is the value of $SS_{between\ treatments}$?
    a. 10
    b. 40
    c. 50
    d. 300

13. For these data, what is the SS value for factor A (SS_A)?
    a. 0
    b. 50
    c. 125
    d. 250

14. If the mean and variance are computed for each sample in an independent-measures two-factor experiment, then which of the following types of sample data will tend to produce large F-ratios for the two-factor ANOVA?
    a. large differences between sample means and small sample variances
    b. large differences between sample means and large sample variances
    c. small differences between sample means and small sample variances
    d. small differences between sample means and large sample variances

15. In a two-factor experiment with 2 levels of factor A and 2 levels of factor B, three of the treatment means are essentially identical and one is substantially different from the others. What result(s) would be produced by this pattern of treatment means?
    a. a main effect for factor A
    b. a main effect for factor B
    c. an interaction between A and B
    d. The pattern would produce main effects for both A and B, and an interaction.

Other Questions

1. Use a two-factor analysis of variance to evaluate the following data from an independent-measures experimental design using n = 5 subjects for each treatment condition (each cell). Use $\alpha = .05$ for all tests.

	B1	B2	B3
A1	M = 1 T = 5 SS = 15	M = 1 T = 5 SS = 15	M = 4 T = 20 SS = 25
A2	M = 1 T = 5 SS = 15	M = 3 T = 15 SS = 25	M = 8 T = 40 SS = 25

$N = 30$
$G = 90$
$\Sigma X^2 = 580$

2. The results from a two-factor research study with 2 levels of factor A, 3 levels of factor B, and n = 5 subjects in each treatment condition were evaluated with an analysis of variance. The results are summarized in the following table. Fill in all missing values.

Source	SS	df	MS	
Between Cells	35	__		
Factor A	__	__	__	$F(1, 24) =$ ____
Factor B	20	__	__	$F(2, 24) =$ ____
AxB	__	__	5	$F(2, 24) =$ ____
Within Cells	__	__	2	
Total	__	__		

ANSWERS TO SELF TEST

True/False Answers

1. True
2. True
3. False. The treatment effect depends on gender.
4. True
5. With six treatment conditions, the study will require 30 subjects.
6. True
7. True
8. False. A two-factor study requires 3 hypothesis tests (2 main effects and 1 interaction).
9. False. The significance of the interaction is completely independent of the significance of the main effects.
10. True

Multiple-Choice Answers

1. d There are 2x3 = 6 different treatments, each with n = 5 participants.
2. d There are 3x4 = 12 different treatment conditions.
3. b df = (3 – 1) and 6(4).
4. d The significance of the interaction is independent of the main effects.
5. b $SS_B = 40$, $SS_{AxB} = 32$, $SS_{within} = 96$, and F = 16/4.
6. b Eta squared = SS_A divided by the sum of SS_A and SS_{within}.
7. d The significance of the interaction is independent of the main effects..
8. a SS between treatments is analyzed, not used to compute a MS..
9. d Both columns should have the same total – no mean difference.
10. b In the top row the mean increases by 10 (from 20 to 30). If there is no interaction, the bottom row should show the same pattern.
11. c df within = $\Sigma(n - 1)$
12. c The SS is computed using the four T values.
13. a Both rows have T = 50 – there is no mean difference.
14. a Large mean differences lead to large numerators for F, and small variance lead to small denominators.
15. d One different mean would create main effects for both factors and an interaction.

Other Answers

1. The results of the two-factor ANOVA are summarized as follows:

Source	SS	df	MS	
Between Cells	190	5		
Factor A	30	1	30	$F(1, 24) = 6.00$
Factor B	140	2	70	$F(2, 24) = 14.00$
AxB	20	2	10	$F(2, 24) = 2.00$
Within Cells	120	24	5	
Total	310	29		

For df = 1, 24 the critical value is 4.26. The F-ratio for factor A is in the critical region so there is a significant difference among the levels of factor A. With df = 2, 24 the critical value is 3.40. The F-ratio for factor B is in the critical region so there are significant differences among the levels of factor B. The interaction is not significant.

2.

Source	SS	df	MS	
Between Cells	35	5		
Factor A	5	1	5	$F(1, 24) = 2.50$
Factor B	20	2	10	$F(2, 24) = 5.00$
AxB	10	2	5	$F(2, 24) = 2.50$
Within Cells	48	24	2	
Total	83	29		

CHAPTER 16

CORRELATION

CHAPTER SUMMARY

The goals of Chapter 16 are:
1. To introduce the concept of a correlation and what it measures.
2. To demonstrate the Pearson correlation to measure linear relationship.
3. To demonstrate the Spearman correlation to measure relationships for ordinal data.
4. To demonstrate the Point-Biserial Correlation and the Phi-Coefficient.

Correlations: Measuring and Describing Relationships

A correlation is a statistical method used to measure and describe the relationship between two variables. A relationship exists when changes in one variable tend to be accompanied by consistent and predictable changes in the other variable.

A correlation typically evaluates three aspects of the relationship: the direction, the form, and the degree of relationship. The **direction** of the relationship is measured by the sign of the correlation (+ or -). A positive correlation means that the two variables tend to change in the same direction; as one increases, the other also tends to increase. A negative correlation means that the two variables tend to change in opposite directions; as one increases, the other tends to decrease. The most common **form** of relationship is a straight line or linear relationship which is measured by the Pearson correlation. The **degree** of relationship (the strength or consistency of the relationship) is measured by the numerical value of the correlation. A value of 1.00 indicates a perfect relationship and a value of zero indicates no relationship.

To compute a correlation you need two scores, X and Y, for each individual in the sample. The Pearson correlation requires that the scores be numerical values from an interval or ratio scale of measurement. Other correlational methods exist for other scales of measurement.

The Pearson Correlation

The **Pearson correlation** measures the direction and degree of linear (straight line) relationship between two variables. To compute the Pearson correlation, you first measure the variability of X and Y scores separately by computing SS for the scores of each variable (SS_X and SS_Y). Then, the covariability (tendency for X and Y to vary together) is measured by the sum of products (SP). The Pearson correlation is found by computing the ratio,

$$SP/\sqrt{(SS_X)(SS_Y)}$$

Thus the Pearson correlation is comparing the amount of covariability (variation from the relationship between X and Y) to the amount X and Y vary separately. The magnitude of the Pearson correlation ranges from 0 (indicating no linear relationship between X and Y) to 1.00 (indicating a perfect straight-line relationship between X and Y). The correlation can be either positive or negative depending on the direction of the relationship.

The Spearman Correlation

The **Spearman correlation** is used in two general situations: (1) It measures the relationship between two ordinal variables; that is, X and Y both consist of ranks. (2) It measures the consistency of direction of the relationship between two variables. In this case, the two variables must be converted to ranks before the Spearman correlation is computed. The calculation of the Spearman correlation requires:

1. Two variables are observed for each individual.

2. The observations for each variable are rank ordered. Note that the X values and the Y values are ranked separately.

3. After the variables have been ranked, the Spearman correlation is computed by either:
    a. Using the Pearson formula with the ranked data.
    b. Using the special Spearman formula (assuming there are few, if any, tied ranks).

The Point-Biserial Correlation and the Phi Coefficient

The Pearson correlation formula can also be used to measure the relationship between two variables when one or both of the variables is dichotomous. A dichotomous variable is one for which there are exactly two categories: for example, men/women or succeed/fail. With either one or two dichotomous variables the calculation of the correlation precedes as follows:
1. Assign numerical values to the two categories of the dichotomous variable(s). Traditionally, one category is assigned a value of 0 and the other is assigned a value of 1.
2. Use the regular Pearson correlation formula to calculate the correlation.

In situations where one variable is dichotomous and the other consists of regular numerical scores (interval or ratio scale), the resulting correlation is called a **point-biserial correlation**. When both variables are dichotomous, the resulting correlation is called a **phi-coefficient**.

The point-biserial correlation is closely related to the independent-measures t test introduced in Chapter 10. When the data consists of one dichotomous variable and one numerical variable, the dichotomous variable can also be used to separate the individuals into two groups. Then, it is possible to compute a sample mean for the numerical scores in each group. In this case, the independent-measures t test can be used to evaluate the mean difference between groups. If the effect size for the mean difference is measured by computing r^2 (the percentage of variance explained), the value of r^2 will be equal to the value obtained by squaring the point-biserial correlation.

LEARNING OBJECTIVES

1. Understand the Pearson correlation and what aspects of a relationship it measures.

2. Know the uses and limitations of measures of correlation.

3. Be able to compute the Pearson correlation by the regular formula (using either the definitional or computational formula for SP) or by the z-score formula.

4. Be able to use a sample correlation to evaluate a hypothesis about the correlation for the general population.

5. Understand the Spearman correlation and how it differs from the Pearson correlation in terms of the data it uses and the type of relationship it measures.

6. Understand how the Pearson correlation formula can be used to compute a point-biserial correlation or a phi-coefficient to measure the relationship between two variables when one or both variables are dichotomous.

7. Understand how the r value obtained for the point-biserial correlation is related to the r^2 value that measures effect size for the independent-measures t test.

NEW TERMS AND CONCEPTS

The following terms were introduced in this chapter. You should be able to define or describe each term and, where appropriate, describe how each term is related to other terms in the list.

Positive relationship A relationship between two variables where increases in one variable tend to be accompanied by increases in the other variable.

Negative relationship A relationship between two variables where increases in one variable tend to be accompanied by decreases in the other variable.

Perfect relationship A relationship where the actual data points perfectly fit the specific form being measured. For a Pearson correlation, the data points fit perfectly on a straight line.

Sum of products (of deviations) A measure of the degree of covariability between two variables; the degree to which they vary together.

Pearson correlation A measure of the direction and degree of linear relationship between two variables.

Significance of a correlation	A demonstration that a sample correlation so large that it is very unlikely ($p < \alpha$) to have come from a population where the correlation is zero.
Coefficient of determination	The degree to the variability in one variable can be predicted by its relationship with another variable: measured by r^2.
Spearman correlation	A correlation calculated for ordinal data. Also used to measure the consistency of direction for a relationship.
Monotonic relation	A relationship that is consistently one-directional.
Point-bacterial correlation	A correlation between two variables where one of the variables is dichotomous.
Phi-coefficient	A correlation between two variables both of which are dichotomous

NEW FORMULAS

$$SP = \Sigma XY - \frac{(\Sigma X)(\Sigma Y)}{n} \quad \text{or} \quad SP = \Sigma(X - M_X)(Y - M_Y)$$

$$r = \frac{SP}{\sqrt{(SS_X)(SS_Y)}}$$

$$\text{Spearman } r_S = 1 - \frac{6\Sigma D^2}{n(n^2 - 1)}$$

CORRELATION

STEP BY STEP

<u>The Pearson</u> Correlation A researcher has pairs of scores (X and Y values) for a sample of n = 5 subjects. The data are as follows:

Person	X	Y
#1	0	−2
#2	2	−5
#3	8	14
#4	6	3
#5	4	0

Step 1: Sketch a scatter plot of the data and make a preliminary estimate of the correlation.

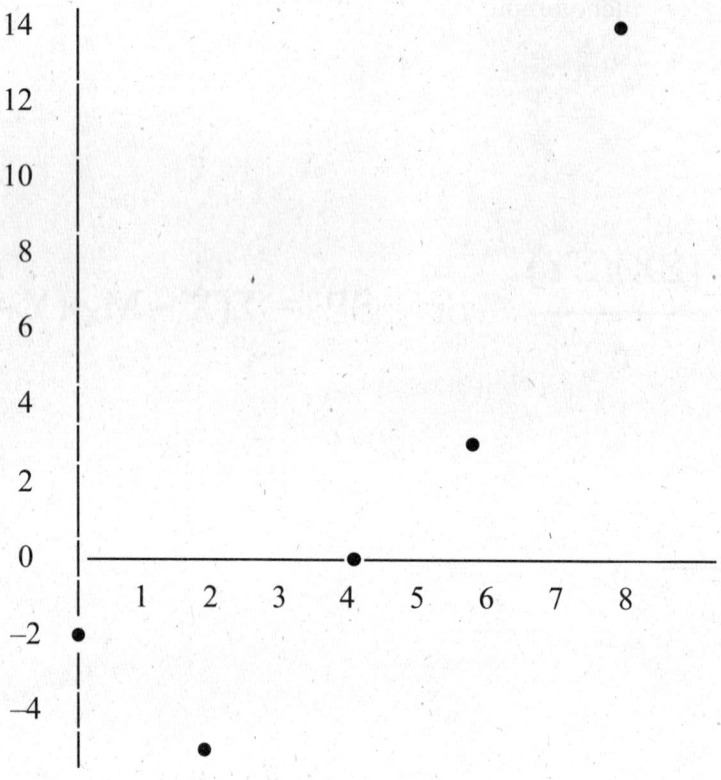

For these data, there appears to be a fairly good, positive correlation - probably around r = +.8 or +.9.

Step 2: To compute the Pearson correlation you must find SS for both X and Y as well as SP. If both sets of scores (X and Y) have means that are whole numbers, then you may use the definitional formulas for SS and SP. Otherwise, it is better to use the computational formulas. Although these data have $M_X = 4$ and $M_Y = 2$, we will demonstrate the computational formulas.

Using one large table, list the X and Y values in the first two columns, then continue with the squared values and the XY products. Find the sum of the numbers in each column. These sums are needed to find SS and SP.

X	Y	X^2	Y^2	XY
0	-2	0	4	0
2	-5	4	25	-10
8	14	64	196	112
6	3	36	9	18
4	0	16	0	0
(totals) 20	10	120	234	120

Use the sums from the table to compute SS for X and Y, and SP.

For X: $SS = \Sigma X^2 - \dfrac{(\Sigma X)^2}{n} = 120 - \dfrac{20^2}{5}$

$= 120 - 80$

$= 40$

For Y: $SS = \Sigma Y^2 - \dfrac{(\Sigma Y)^2}{n} = 234 - \dfrac{10^2}{5}$

$= 234 - 20$

$= 214$

and $SP = \Sigma XY - \dfrac{(\Sigma X)(\Sigma Y)}{n} = 120 - \dfrac{(20)(10)}{5}$

$= 120 - 40$

$= 80$

Step 3: Compute the Pearson correlation and compare the answer with your preliminary estimate from Step 1. For these data,

$$r = \frac{SP}{\sqrt{(SS_X)(SS_Y)}} = \frac{80}{\sqrt{(40)(214)}} = \frac{80}{92.52} = 0.865$$

The obtained correlation, r = +0.865 agrees with our preliminary estimate.

<u>The Spearman Correlation</u> The following example demonstrates the calculation of the Spearman Correlation: The Spearman correlation measures the degree of relationship between two variables that are both measured on ordinal scales. If the original data are from interval or ratio scales, you can rank the scores, then compute the Spearman correlation. The following data will be used to demonstrate the calculation of the Spearman correlation.

X	Y	
5	12	
7	18	(Both X and Y are measured on
2	9	interval scales)
15	14	
10	13	

Step 1: Check that the X values and the Y values consist of ranks (ordinal data). If not, rank the X's and rank the Y's. Caution: Rank X and Y separately.

For these data,

X Score	X Rank	Y Rank	Y Score
5	2	2	12
7	3	5	18
2	1	1	9
15	5	4	14
10	4	3	13

Step 2: To use the special Spearman formula, compute the difference (D) between the X rank and the Y rank for each individual. Also, find the squared difference (D^2) and the sum of the squared differences.

(Note: The signs of the difference scores are unimportant because you are squaring each D.)

X Rank	Y Rank	D	D²
2	2	0	0
3	5	2	4
1	1	0	0
5	4	1	1
4	3	1	1
			6 = ΣD²

Step 3: Substitute ΣD^2 and n in the Spearman formula.

$$r_S = 1 - \frac{6\Sigma D^2}{n(n^2 - 1)}$$

$$= 1 - \frac{6(6)}{5(25 - 1)}$$

$$= 1 - \frac{36}{120}$$

$$= 0.70$$

There is a positive relation between X and Y for these data. The correlation is fairly high (near 1.00) which indicates a very consistent positive relation.

HINTS AND CAUTIONS

1. Remember, a correlation of –0.80 is just as strong as a correlation of +0.80. The sign indicates the direction of the relationship, not its magnitude.

2. Remember that n refers to the number of individuals (pairs of scores).

3. The special formula for the Spearman correlation often causes trouble. Remember, the value of the fraction is computed separately and then subtracted from 1.00. The 1 is not a part of the fraction.

4. When computing either the point-biserial correlation or the phi-coefficient, the numerical values are assigned arbitrarily to the two dichotomous categories. As a result, the sign of the correlation (+ or n –) is also arbitrary and meaningless.

SELF TEST

True/False Questions

1. For a negative correlation, as the X values decrease from one person to another, the Y values also tend to decrease.

2. A Pearson correlation of r = –0.90 indicates that the data points are clustered close around a line that slopes down to the right.

3. The value for a correlation can never be greater than 1.00.

4. The value of sum of products (SP) can never be less than zero.

5. A researcher observing children on a school playground noticed that the number of aggressive acts tends to increase as the temperature increases. This is an example of a positive correlation.

6. One individual with an X or Y value that is substantially different from the X or Y values for the rest of the group can have a dramatic impact on the value of the Pearson correlation.

7. For a two-tailed hypothesis test evaluating the significance of a Pearson correlation, the null hypothesis states that the correlation for the population is zero.

8. The hypothesis test evaluating the significance of a Pearson correlation from a sample of n = 15 would have df = 14.

9. The Spearman correlation is used to measure the relationship when both variables have been measured on an ordinal scale (both are ranks).

10. The phi-coefficient is used to measure relationships for data that would also be appropriate for an independent-measures t test.

Multiple-Choice Questions

1. During the winter in New York, the lower the temperature drops, the higher the demand for energy. This is an example of
    a. a positive relationship.
    b. a negative relationship.
    c. a cause-effect relationship.
    d. all of the above.

2. What pattern would appear in a graph of a Pearson correlation of r = +0.80?
    a. points tightly clustered around a line that slopes up to the right.
    b. points tightly clustered around a line that slopes down to the right.
    c. points widely scattered around a line that slopes up to the right.
    d. points widely scattered around a line that slopes down to the right.

3. The scatter plot for a set of X and Y values shows the data points clustered in a nearly perfect circle. For these data, what is the most likely value for the Pearson correlation?
    a. r near 0
    b. r near –1.00
    c. r near +1.00

4. Which of the following Pearson correlations shows the greatest strength or consistency of relationship?
    a. –0.90
    b. +0.74
    c. +0.85
    d. –0.33

5. For which of the following Pearson correlations would the data points be clustered most closely around a straight line?
    a. r = –0.10
    b. r = +0.40
    c. r = –0.70
    d. There is no relationship between the correlation and how close the points are to a straight line.

6. A set of n = 10 pairs of scores has $\Sigma X = 20$, $\Sigma Y = 30$, and $\Sigma XY = 74$. What is the value of SP for these data?
    a. 74
    b. 24
    c. 14
    d. –14

7. A set of n = 15 pairs of scores (X and Y values) produces a correlation of r = 0.40. If each of the X values is multiplied by 2 and the correlation is computed for the new scores, what value will be obtained for the new correlation?
    a. r = 0.20
    b. r = 0.40
    c. r = 0.80
    d. cannot be determined without knowing all the X and Y scores

8. A set of n = 5 pairs of X and Y scores has $\Sigma X = 15$, $\Sigma Y = 5$, and $\Sigma XY = 10$. For these data, the value of SP is
   a. –5
   b. 5
   c. 10
   d. 25

9. The value of the Pearson correlation will be negative if
   a. the value of SS_X is negative.
   b. the value of SS_Y is negative.
   c. either SS_X or SS_Y is negative, but not both.
   d. SP is negative.

10. When the Pearson correlation is computed for data measured on an ordinal scale (ranks), the result is know as
    a. the Spearman correlation.
    b. the point-biserial correlation.
    c. the phi coefficient.
    d. It is still called a Pearson correlation.

11. Under what circumstances is the phi-coefficient used?
    a. When one variable consists of ranks and the other is regular, numerical scores.
    b. When both variables consists of ranks.
    c. When both X and Y are dichotomous variables
    d. When one variable is dichotomous and the other is regular, numerical scores.

12. A set of n = 15 pairs of scores (X and Y values) has $SS_X = 4$, $SS_Y = 25$, and SP = 6. The Pearson correlation for these data is
    a. 6/100
    b. 6/10
    c. 6/(100/15)
    d. 6/(10/√15)

13. A correlation is computed for a sample of n = 18 pairs of X and Y values. What correlations are statistically significant with $\alpha = .05$, two tails.
    a. correlations between 0.468 and –0.468
    b. correlations greater than or equal to 0.468 and correlation less than or equal to –0.468
    c. correlations between 0.456 and –0.456
    d. correlations greater than or equal to 0.456 and correlation less than or equal to –0.456

14. If the following seven scores are ranked from smallest (#1) to largest (#7) what rank should be assigned to a score of X = 6? Scores: 1, 1, 3, 6, 6, 6, 9
    a. 3
    b. 4
    c. 5
    d. 6

15. Under what circumstances is the point-biserial correlation used?
    a. in the same circumstances when a repeated-measures t test would be used
    b. in the same circumstances when an independent-measures t test would be used
    c. when both X and Y are dichotomous variables
    d. when both X and Y are measured on an ordinal scale (ranks)

Other Questions

1. Compute SP for the following sets of data. You should find that the definitional formula works well with Set I because both means are whole numbers. However, the computational formula is better with Set II where the means are fractions.

Data Set I		Data Set II	
X	Y	X	Y
1	5	1	0
5	2	4	4
6	9	3	1
15	20	2	1
8	4		

2. Compute the Pearson correlation for the following set of scores:

X	Y
2	2
3	7
4	6
2	4
4	6

3. Suppose that a sample of n = 42 pairs of X and Y scores yields a Pearson correlation of r = +0.40. Does this sample provide sufficient evidence to conclude that a significant correlation exists in the population? Test at the .05 level of significance, two tails.

4. Compute the Spearman correlation for each of the following sets of data. (Note that you will need to rank order the X and Y values for the data in Set 2.)

Set 1 X and Y measured on ordinal scales		Set 2 X and Y measured on interval scales	
X	Y	X	Y
2	5	1	5
4	1	5	2
3	2	6	9
1	4	15	20
5	3	8	4

5. Compute the point-biserial correlation for the following set of data. (Note that you will need to convert the Male/Female categories into 0/1 scores.

X	Y
Male	3
Female	7
Female	4
Male	7
Female	5
Male	2
Male	4
Female	8

ANSWERS TO SELF TEST

True/False Answers

1. False. For a negative correlation, decreases in one variable tend to be accompanied by increases in the other variable.
2. True
3. True
4. False. The value of SP can be either positive or negative.
5. True
6. True
7. True
8. False. $df = n - 2 = 13$.
9. False. The value of the Spearman correlation can range in value from +1.00 to -1.00.
10. False. The point-biserial correlation uses data that would also be appropriate for an independent-measures t test.

Multiple-Choice Answers

1. b Temperature and energy demand change in opposite directions.
2. a This is a consistent positive relationship.
3. a With no positive and no negative trend the correlation is near zero.
4. a The larger the correlation (independent of sign), the stronger the relationship.
5. c The larger the numerical value of the correlation (independent of sign) the closer the points are clustered around a straight line.
6. c SP = 74 – 20(30)/10
7. b Multiplying by a constant will not change the pattern in the scatter plot. The correlation will not change.
8. a SP = 10 – 5(15)/5
9. d The sign of SP controls the sign of the correlation. SS is always positive.
10. a This is the method for computing a Spearman correlation.
11. c Phi is used to measure the relationship between two dichotomous variables.
12. b r = SP divided by the square root of (SS_X + SS_Y)
13. b With df = 16 the critical value is ±0.468.
14. c The three 6s are tied for 4th, 5th, and 6th.
15. b The point-biserial correlation is used when one variable is dichotomous and the second consists of numerical scores; that is, two groups of scores.

Other Answers

1. For data set I, SP = 121. For data set II, SP = 6.

2. a. SS_X = 4, SS_Y = 16, and SP = 6. The correlation is r = 0.75.
   b. $\hat{Y}$ = 1.5X + 0.5

3. The null hypothesis states that there is no relationship in the population. H_0: ρ = 0. With n = 42, the correlation must be greater than 0.304 to be significant. This sample correlation is sufficient to conclude that there is a significant correlation in the population.

4. For data set 1, the Spearman correlation is r_S = –0.60. After ranking the scores in data set 2, the Spearman correlation is r_S = +0.50.

5. Using Male = 0 and Female = 1, SS = 2, SS = 32, and SP = 4.5 and the point-biserial correlation is r = 0.563.

CHAPTER 17

INTRODUCTION TO REGRESSION

CHAPTER SUMMARY

The goals of Chapter 17 are:
1. To introduce the process of linear regression.
2. To demonstrate the analysis of regression to evaluate the significance of a linear regression equation.
3. To introduce the process of multiple regression with two predictor variables.
4. To demonstrate analysis of regression as it applies to a multiple regression equation.
5. To introduce the concept of standard error of estimate as a measure of the average error (or discrepancy) between the actual Y values and the values predicted by the regression equation.

Introduction to Linear Regression

The Pearson correlation measures the degree to which a set of data points form a straight line relationship. **Regression** is a statistical procedure that determines the equation for the straight line that best fits a specific set of data.

Any straight line can be represented by an equation of the form $Y = bX + a$, where b and a are constants. The value of b is called the slope constant and determines the direction and degree to which the line is tilted. The value of a is called the Y-intercept and determines the point where the line crosses the Y-axis.

How well a set of data points fits a straight line can be measured by calculating the distance between the data points and the line. The total error between the data points and the line is obtained by squaring each distance and then summing the squared values. The regression equation is designed to produce the minimum sum of squared errors. The equation for the regression line is

$$\hat{Y} = bX + a \quad \text{where} \quad b = \frac{SP}{SS_X} \quad \text{and} \quad a = M_Y - bM_X$$

The ability of the regression equation to accurately predict the Y values is measured by first computing the proportion of the Y-score variability that is predicted by the regression equation and the proportion that is not predicted.

Predicted variability = $SS_{regression} = r^2 SS_Y$ with $df = 1$

Unpredicted variability = $SS_{residual} = (1 - r^2)SS_Y = \Sigma(Y - \hat{Y})^2$ with $df = n - 2$

The unpredicted variability can be used to compute the standard error of estimate which is a measure of the average distance between the actual Y values and the predicted Y values.

$$\text{standard error of estimate} = \sqrt{\frac{SS_{residual}}{df}}$$

Finally, the overall significance of the regression equation can be evaluated by computing an F-ratio. A significant F-ratio indicates that the equation predicts a significant portion of the variability in the Y scores (more than would be expected by chance alone). To compute the F-ratio, you first calculate a variance or MS for the predicted variability and for the unpredicted variability:

$$MS_{regressiion} = \frac{SS_{regression}}{df_{regression}} \quad \text{and} \quad MS_{residuals} = \frac{SS_{residuals}}{df_{residuals}}$$

The F-ratio is

$$F = \frac{MS_{regressiion}}{MS_{residuals}} \quad \text{with } df = 1, n - 2$$

Introduction to Multiple Regression with Two Predictor Variables

In the same way that linear regression produces an equation that uses values of X to predict values of Y, multiple regression produces an equation that uses two different variables (X_1 and X_2) to predict values of Y. The equation is determined by a least squared error solution that minimizes the squared distances between the actual Y values and the predicted Y values. For two predictor variables, the general form of the multiple regression equation is: $\hat{Y} = b_1X_1 + b_2X_2 + a$

Formulas for computing b_1, b_2, and a are presented in the New Equations section that follows.

The ability of the multiple regression equation to accurately predict the Y values is measured by first computing the proportion of the Y-score variability that is predicted by the regression equation and the proportion that is not predicted.

Predicted variability = $SS_{regression}$ = $R^2 SS_Y$ with $df = 2$

Unpredicted variability = $SS_{resicual}$ = $(1 - R^2)SS_Y$ = $\Sigma(Y - \hat{Y})^2$ with $df = n - 3$

$$\text{where } R^2 = \frac{b_1 SP_{X1Y} + b_2 SP_{X2Y}}{SS_Y}$$

As with linear regression, the unpredicted variability (SS and df) can be used to compute a standard error of estimate that measures the standard distance between the actual Y values and the predicted values. In addition, the overall significance of the multiple regression equation can be evaluated with an F-ratio:

$$F = \frac{MS_{regressiion}}{MS_{residuals}} \quad \text{with df} = 2, n - 3$$

where each MS = SS/df

Partial Correlation

A partial correlation measures the relationship between two variables (X and Y) while eliminating the influence of a third variable (Z). Partial correlations are used to reveal the real, underlying relationship between two variables when researchers suspect that the apparent relation may be distorted by a third variable. For example, there probably is no underlying relationship between weight and mathematics skill for elementary school children. However, both of these variables are positively related to age: Older children weigh more and, because they have spent more years in school, have higher mathematics skills. As a result, weight and mathematics skill will show a positive correlation for a sample of children that includes several different ages. A partial correlation between weight and mathematics skill, holding age constant, would eliminate the influence of age and show the true correlation which is near zero.

LEARNING OBJECTIVES

1. Recognize the general form of a linear equation and be able to identify its slope and Y-intercept.

2. Be able to compute the linear regression equation for a set of data.

3. Be able to use the regression equation to compute a predicted value of Y for any given value of X.

4. Be able to compute the multiple regression equation for a set of data with two predictor variables.

5. Be able to compute the standard error of estimate for either a linear regression equation or a multiple regression equation.

6. Be able to evaluate the significance of either a linear regression or a multiple regression equation by computing the appropriate F-ratio.

7. Understand the concept of a partial correlation.

NEW TERMS AND CONCEPTS

The following terms were introduced in this chapter. You should be able to define or describe each term and, where appropriate, describe how each term is related to other terms in the list.

Linear relationship	A relationship between two variables where a specific increase in one variable is always accompanied by a specific increase (or decrease) in the other variable.
Linear equation	An equation of the form $Y = bX + a$ expressing the relationship between two variables X and Y.
Slope	The amount of change in Y for each 1-point increase in X. The value of b in the linear equation.
Y-intercept	The value of Y when $X = 0$. In the linear equation, the value of a.
Regression equation for Y	The equation for the best-fitting straight line to describe the relationship between X and Y.
Multiple regression equation	The equation producing the most accurate predictions for Y based on two predictor variables. Accuracy is defined as having the least squared error between the actual Y values and the predicted values.
Predicted variability of Y	The proportion of the variability for the Y scores that is predicted by the regression equation. Determined by r^2 for linear regression or R^2 for multiple regression.
Unpredicted variability of Y	The proportion of the variability for the Y scores that is not predicted by the regression equation. (Also known as the residual variability.) Determined by $1 - r^2$ for linear regression or $1 - R^2$ for multiple regression.
Standard error of estimate	A measure of the average distance between the actual Y values and the predicted values from the regression equation.
Analysis of regression	Evaluating the significance of a regression equation by computing an F-ratio comparing the predicted variance (MS) in the numerator and the unpredicted variance (MS) in the denominator.

INTRODUCTION TO REGRESSION

NEW FORMULAS

For linear regression:

$$\hat{Y} = bX + a \qquad b = \frac{SP}{SS_X} \qquad a = M_Y - bM_X$$

$$SS_{regression} = r^2 SS_Y \qquad df = 1 \qquad MS_{regression} = SS/df$$

$$SS_{residual} = (1 - r^2)SS_Y \qquad df = n - 2 \qquad MS_{residual} = SS/df$$

$$F = \frac{MS_{regression}}{MS_{residual}}$$

For multiple regression:

$$\hat{Y} = b_1 X_1 + b_2 X_2 + a$$

$$b_1 = \frac{(SP_{X1Y})(SS_{X2}) - (SP_{X1X2})(SP_{X2Y})}{(SS_{X1})(SS_{X2}) - (SP_{X1X2})^2}$$

$$b_2 = \frac{(SP_{X2Y})(SS_{X1}) - (SP_{X1X2})(SP_{X1Y})}{(SS_{X1})(SS_{X2}) - (SP_{X1X2})^2}$$

$$a = M_Y - b_1 M_{X1} - b_2 M_{X2}$$

$$R^2 = \frac{b_1 SP_{X1Y} + b_2 SP_{X2Y}}{SS_Y}$$

$$SS_{regression} = R^2 SS_Y \quad df = 2 \quad MS_{regression} = SS/df$$

$$SS_{residual} = (1 - R^2)SS_Y \quad df = n - 3 \quad MS_{residual} = SS/df$$

$$F = \frac{MS_{regression}}{MS_{residual}}$$

The partial correlation between X and Y, holding Z constant:

$$r_{XY \cdot Z} = \frac{r_{XY} - (r_{XZ}\, r_{YZ})}{\sqrt{(1 - r_{XZ}^2)(1 - r_{YZ}^2)}}$$

STEP BY STEP

<u>Linear Regression</u>   A researcher has pairs of scores (X and Y values) for a sample of n = 5 subjects. The data are as follows:

Person	X	Y
#1	0	−2
#2	2	−5
#3	8	14
#4	6	3
#5	4	0

Step 1: Sketch a scatterplot of the data and make a preliminary estimate of the correlation. Also, sketch a line through the middle of the data points and note the slope and Y-intercept of the line corresponding to the regression equation.

INTRODUCTION TO REGRESSION

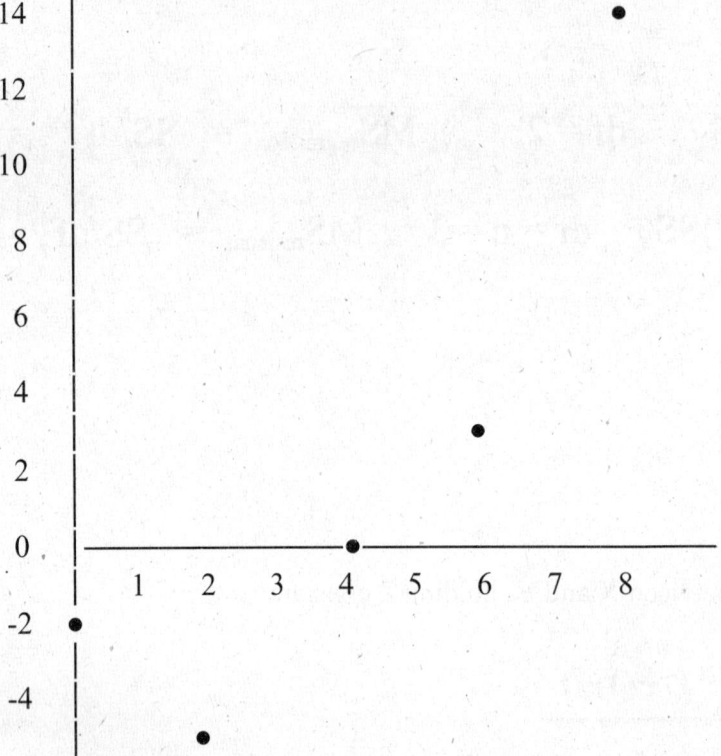

For these data, there appears to be a fairly good, positive correlation - probably around r = +.7 or +.8. The line has a positive slope and appears to intersect the Y-axis about 5 points below zero.

Step 2: To find the regression equation you must find SS_X and the value for SP, as well as the mean for the X scores and the Y scores. Because the same values are used to compute the correlation, and because the correlation is used to help evaluate the significance of the regression equation, we will compute the correlation as well as the regression equation. For these data, M_X and M_Y are both whole numbers ($M_X = 4$ and $M_Y = 2$) so you can use the definitional formulas for SS and SP. However, we will demonstrate the calculations with the computational formulas.

Using one large table, list the X and Y values in the first two columns, then continue with the squared values and the XY products. Find the sum of the numbers in each column. These sums are needed to find SS and SP.

	X	Y	X^2	Y^2	XY
	0	−2	0	4	0
	2	−5	4	25	-10
	8	14	64	196	112
	6	3	36	9	18
	4	0	16	0	0
(totals)	20	10	120	234	120

Use the sums from the table to compute SS for X and Y, and SP.

For X: $SS = \Sigma X^2 - \dfrac{(\Sigma X)^2}{n} = 120 - \dfrac{20^2}{5}$

$= 120 - 80$

$= 40$

For Y: $SS = \Sigma Y^2 - \dfrac{(\Sigma Y)^2}{n} = 234 - \dfrac{10^2}{5}$

$= 234 - 20$

$= 214$

and $SP = \Sigma XY - \dfrac{(\Sigma X)(\Sigma Y)}{n} = 120 - \dfrac{(20)(10)}{5}$

$= 120 - 40$

$= 80$

Step 3: Compute the Pearson correlation and the regression equation For these data,

$$r = \dfrac{SP}{\sqrt{(SS_X)(SS_Y)}} = \dfrac{80}{\sqrt{(40)(214)}} = \dfrac{80}{92.52} = 0.865$$

Note that the obtained correlation, r = +0.865, agrees with our preliminary estimate.

The general form of the regression equation is, $\hat{Y} = bX + a$. For these data,

$$b = \dfrac{SP}{SS_X} = \dfrac{80}{40} = 2$$

and $a = a = M_Y - bM_X$

$= 2 - 2(4)$

$= -6$

The regression equation is: $\hat{Y} = 2X - 6$

INTRODUCTION TO REGRESSION

Both the slope constant (b = +2) and the Y-intercept (a = –6) agree with our preliminary estimates.

Analysis of Regression The significance of the regression equation can be evaluated using an analysis of regression. The null hypothesis says that there is no consistent relationship between X and Y in the population and the regression equation does not predict a significant portion of the variance for the Y scores.

The analysis of regression is based on an F-ratio comparing the variance predicted by the equation and the unpredicted, or residual variance. The predicted portion of the variance is determined by r^2 and the unpredicted portion by $1 - r^2$. For the data we have been considering, r = 0.865 and SS_Y = 214. Using these values, we obtain

$$SS_{regression} = r^2 SS_Y = (0.865)^2(214) = 0.748(214) = 160.07$$

$$SS_{residual} = (1 - r^2)SS_Y = (1 - 0.748)(214) = 0.252(214) = 53.93$$

$$MS_{regression} = \frac{SS_{regression}}{df} = \frac{160.07}{1} = 160.07$$

$$MS_{residual} = \frac{SS_{residual}}{df} = \frac{53.93}{n-2} = \frac{53.93}{5-2} = 17.98$$

and the F-ratio is, $F = \dfrac{MS_{regression}}{MS_{residual}} = \dfrac{160.07}{17.98} = 8.90$

With df = 2, 3 and α = .05, the critical value is 9.55. The decision is to fail to reject the null hypothesis. Our conclusion is that the regression equation does not predict a significant portion of the variance for the Y scores.

HINTS AND CAUTIONS

1. Although linear regression and multiple regression can be conducted without first computing a correlation, it is usually helpful to calculate either the Pearson correlation or R (for multiple regression), especially if you plan to evaluate the significance of the regression equation.

2. The formula for the Y-intercept (a) in the regression equation may be easier to remember if you note that this formula simply guarantees that the point defined by (M_X, M_Y) is on the regression line. Thus, when X = M_X, the predicted Y score will be M_Y. In the equation, $M_Y = bM_X + a$. Solving for this equation for the value of a yields, $a = M_Y - bM_X$. Similarly, the multiple regression equation is structured so that using the means for X_1 and X_2 in the equation will generate a predicted Y value of M_Y.

3. Remember that a regression equation should not be used to predict values outside the range of the original data.

SELF TEST

True/False Questions

1. The line defined by the $Y = -3X + 6$ slopes up to the right.

2. For linear regression, if $X = M_X$, then the predicted Y value is M_Y.

3. For a regression equation with a positive slope, if the X value is above the mean for the X scores, then the predicted Y value will be above the mean for the Y scores.

4. A set of X and Y scores has $SS_X = 10$, $SS_Y = 36$, and $SP = 20$. The regression equation for these scores will have a slope constant of 2.

5. A set of $n = 6$ pairs of X and Y values has a Pearson correlation of $r = 0.80$ and $SS_Y = 100$. The standard error of estimate for the regression equation is $\sqrt{36/4} = 3$ points.

6. The F-ratio evaluating the significance of a linear regression equation based on $n = 10$ pairs of X and Y scores has $df = 1, 8$.

7. The value of $SS_{residual}$ measures the total squared distance between the actual Y values and the Y values predicted by the regression equation.

8. The proportion of SS_Y that is not predicted by the regression equation is equal to r^2.

9. For multiple regression, the value of R^2 is computed to measure the proportion of the variability for the Y scores that is predicted by the regression equation.

10. In a multiple regression equation with two predictor variables, if b_1 is larger than b_2, then you can conclude that X_1 is a better predictor than is X_2.

Multiple-Choice Questions

1. A set of $n = 15$ pairs of X and Y scores has $SS_X = 10$, $SS_Y = 40$, and $SP = 30$. What is the slope for the regression equation for predicting Y from X?
    a. 10/30
    b. 10/40
    c. 40/10
    d. 30/10

2. A set of X and Y scores has a regression equation with a slope of $b = 4$. If the mean for the X values is $M_X = 2$ and the mean for the Y values is $M_Y = 6$, then what is the Y-intercept value for the regression equation?
    a. –2
    b. –22
    c. 4
    d. –4

3. The standardized form of the regression equation is $z_Y = (beta)z_X$. In the equation, what is the value of beta?
    a. Beta is equal to the correlation between X and Y.
    b. Beta is equal to SP/ SS_X.
    c. Beta is equal to r/(n – 1)
    d. Beta is equal to r/n

4. The distance between the Y value in the data and the Y value predicted from the regression equation is know as the *residual*. What is the value for the sum of the squared residuals?
    a. $SS_{residual} = r^2(SS_X)$
    b. $SS_{residual} = (1 – r^2)(SS_X)$
    c. $SS_{residual} = r^2(SS_Y)$
    d. $SS_{residual} = (1 – r^2)(SS_Y)$

5. For a linear regression equation, what is the relationship between the standard error of estimate and the correlation?
    a. The larger the value of the correlation (either positive or negative), the smaller the standard error of estimate.
    b. The smaller the value of the correlation (either positive or negative), the smaller the standard error of estimate.
    c. The standard error of estimate is not related to the size of the correlation.

6. One use of regression equation is to increase the accuracy of predicting Y scores. How much of the variability in the Y scores is predictable from the regression equation?
    a. r^2
    b. $1 – r^2$
    c. r
    d. $1 – r$

7. For the regression equation, $\hat{Y} = –2X + 6$, what can be determined about the correlation between X and Y?
    a. The correlation definitely will be positive.
    b. The correlation definitely will be negative.
    c. The correlation will be relatively large and positive
    d. The correlation will be relatively large and negative.

8. For the regression equation, $\hat{Y} = –2X + 6$, if the X value is above the mean (positive deviation), then what can be determined about the predicted Y value?
    a. The predicted Y value will be above the mean for the Y scores.
    b. The predicted Y value will be below the mean for the Y scores.
    c. It is impossible to determine where the predicted Y value will be.

9. An analysis of regression is used to test the significance of a linear regression equation based on a sample of n = 20 individuals. What are the df values for the F-ratio?
   a. 1, 18
   b. 1, 19
   c. 2, 18
   d. 2, 19

10. A linear regression equation has a slope constant of b = 2 and a Y-intercept of a = 3. What is the predicted value of Y for X = 8?
   a. $\hat{Y} = 5$
   b. $\hat{Y} = 19$
   c. $\hat{Y} = 26$
   d. cannot be determined without additional information

11. For a linear regression equation, the sum of the squared residuals can be computed directly by finding the difference between each Y and its predicted Y, then squaring the difference and adding the squared values. An alternative procedure is to calculate
   a. $r^2 SS_Y$
   b. $(1 - r^2)SS_Y$
   c. $(1 + r^2)SS_Y$
   d. SS_Y/SS_X

12. A multiple regression equation is computed for a sample of n = 20 sets of X_1, X_2, and Y scores. If the significance of the regression equation is evaluated using an F-ratio, then the ratio would have degrees of freedom equal to
   a. df = 1, 18
   b. df = 1, 19
   c. df = 2, 17
   d. df = 2, 18

13. The Pearson correlation between X_1 and Y is r = 0.40 and SS_Y = 100. When a second variable, X_2, is added to the regression equation, we obtain R^2 = 0.25. How much additional variance is predicted by adding the second variable compared to using X_1 alone?
   a. 9 points
   b. 15 points
   c. 25 points
   d. 40 points

14. A researcher who is evaluating the significance of a multiple regression equation with two predictor variables, obtains an F-ratio with df = 2, 18. How many individuals were needed to produce the sample data used to find the regression equation?
   a. 18
   b. 19
   c. 20
   d. 21

15. The criteria for "best fitting" equation for either linear regression or multiple regression, is the equation that produces the smallest value for
    a. $\Sigma(Y - \hat{Y})$
    b. $\Sigma(Y - \hat{Y})^2$
    c. r^2 or R^2
    d. $SS_{regression}$

Other Questions

1. For the following set of scores
    a. Compute the Pearson correlation.
    b. Find the regression equation for predicting Y from X.

X	Y
0	16
1	6
2	9
3	0
4	9

2. Find the regression equation for the following set of data.

X	Y
4	1
7	16
3	4
5	7
6	7

3. Suppose that a sample of n = 12 pairs of X and Y scores has $SS_Y = 90$ and a Pearson correlation of r = +0.40.
    a. What proportion of the variance for the Y scores is predicted by the regression equation?
    b. Does the regression equation predict a significant portion of the variance? Test with $\alpha = .05$.

4. A multiple regression equation with two predictors is calculated using the scores from a sample of n = 15 people. If the scores have $SS_Y = 80$ and $R^2 = .25$, then
    a. What proportion of the variance for the Y scores is predicted by the regression equation?
    b. Does the regression equation predict a significant portion of the variance? Test with $\alpha = .05$.

ANSWERS TO SELF TEST

True/False Answers

1. False. For a negative correlation, decreases in one variable tend to be accompanied by increases in the other variable.
2. True
3. True
4. True
5. True
6. True
7. True
8. False. The unpredicted portion is $1 - r^2$
9. True
10. False. The values of b_1 and b_2 do not indicate the amount of predicted variance.

Multiple-Choice Answers

1. d The slope is SP divided by SS_X.
2. c $SS_{residual}$ is determined by $1 - r^2$
3. a $a = M_Y - bM_X$
4. d $(1 - r^2)$ is the proportion of SS_Y that is unpredicted.
5. a The standard error is determined by $(1 - r^2)$
6. a r^2 is the proportion of SS_Y that is predicted.
7. b A negative slope constant indicates a negative correlation.
8. b With a negative slope, positive deviations for X will predict negative deviations for Y.
9. a For linear regression, $df = 1, n - 2$
10. b $2(8) + 3 = 19$
11. b $SS_{residual}$ is determined by $(1 - r^2)SS_Y$
12. c For multiple regression, $df_{regression} = 2$ and $df_{residual} = n - 3$
13. a The multiple regression equation predicts $R^2(SS_Y) = 25$ points and the linear regression predicts $r^2(SS_Y) = 16$ points. The difference is 9 points.
14. d $df_{residual} = n - 3 = 18$. $n = 21$
15. b The regression equation is based on the least squared error between the actual Y values and the values predicted by the equation.

INTRODUCTION TO REGRESSION

Other Answers

1. a. $SS_X = 10$, $SS_Y = 134$, and $SP = -20$. The Pearson correlation is $r = -0.546$.
    b. The regression equation is, $\hat{Y} = -2X + 12$.

2. For these data, $SS_X = 10$ and $SP = 30$. The regression equation is, $\hat{Y} = 3X - 8$.

3. a. The predicted portion is $r^2 = 0.16$ (or 16%)
    b. $SS_{regression} = r^2(SS_Y) = 0.16(90) = 14.4$ with $df = 1$. $SS_{residual} = (1 - r^2)(SS_Y) = 0.84(90) = 75.6$ with $df = 10$. The F-ratio is $F = 14.4/7.56 = 1.90$. With $df = 1, 10$, this is not significant.

4. a. The predicted portion is $R^2 = 0.25$ (or 25%)
    b. $SS_{regression} = R^2(SS_Y) = 0.25(80) = 20$ with $df = 2$. $SS_{residual} = (1 - R^2)(SS_Y) = 0.75(80) = 60$ with $df = n - 3 = 12$. The F-ratio is $F = 10/5 = 2.00$. With $df = 2, 12$, this is not significant.

CHAPTER 18

HYPOTHESIS TESTS WITH CHI-SQUARE

CHAPTER SUMMARY

The goals of Chapter 18 are:
1. To introduce the distinction between parametric and nonparametric statistical tests.
2. To demonstrate the chi-square test for goodness of fit.
3. To demonstrate the chi-square test for independence.
4. To demonstrate the measurement of effect size for the test for independence.

Parametric and Nonparametric Tests
    Chapter 18 introduces two **non-parametric hypothesis tests** using the chi-square statistic: the chi-square test for goodness of fit and the chi-square test for independence. The term "non-parametric" refers to the fact that the chi-square tests do not require assumptions about population parameters nor do they test hypotheses about population parameters. Previous examples of hypothesis tests, such as the t tests and analysis of variance, are **parametric tests** and they do include assumptions about parameters and hypotheses about parameters. The most obvious difference between the chi-square tests and the other hypothesis tests we have considered (t and ANOVA) is the nature of the data. For chi-square, the data are frequencies rather than numerical scores.

The Chi-Square Test for Goodness-of-Fit
    The **chi-square test for goodness-of-fit** uses frequency data from a sample to test hypotheses about the shape or proportions of a population. Each individual in the sample is classified into one category on the scale of measurement. The data, called **observed frequencies**, simply count how many individuals from the sample are in each category. The null hypothesis specifies the proportion of the population that should be in each category. The proportions from the null hypothesis are used to compute **expected frequencies** that describe how the sample would appear if it were in perfect agreement with the null hypothesis.

The Chi-Square Test for Independence
    The second chi-square test, the **chi-square test for independence**, can be used and interpreted in two different ways:
1. Testing hypotheses about the relationship between two variables in a population, or
2. Testing hypotheses about differences between proportions for two or more populations.

    Although the two versions of the test for independence appear to be different, they are equivalent and they are interchangeable. The first version of the test emphasizes the relationship between chi-square and a correlation, because both procedures examine the relationship between two variables. The second version of the test emphasizes the relationship between chi-square and an independent-measures t test (or ANOVA) because both tests use data from two (or more) samples to test hypotheses about the difference between two (or more) populations.

The first version of the chi-square test for independence views the data as one sample in which each individual is classified on two different variables. The data are usually presented in a matrix with the categories for one variable defining the rows and the categories of the second variable defining the columns. The data, called **observed frequencies**, simply show how many individuals from the sample are in each cell of the matrix. The null hypothesis for this test states that there is no relationship between the two variables; that is, the two variables are independent.

The second version of the test for independence views the data as two (or more) separate samples representing the different populations being compared. The same variable is measured for each sample by classifying individual subjects into categories of the variable. The data are presented in a matrix with the different samples defining the rows and the categories of the variable defining the columns. The data, again called **observed frequencies**, show how many individuals are in each cell of the matrix. The null hypothesis for this test states that the proportions (the distribution across categories) are the same for all of the populations.

Both chi-square tests use the same statistic. The calculation of the chi-square statistic requires two steps:

1. The null hypothesis is used to construct an idealized sample distribution of **expected frequencies** that describes how the sample would look if the data were in perfect agreement with the null hypothesis.

   For the goodness of fit test, the expected frequency for each category is obtained by
   $$\text{expected frequency} = f_e = pn$$

   (p is the proportion from the null hypothesis and n is the size of the sample)
   For the test for independence, the expected frequency for each cell in the matrix is obtained by
   $$\text{expected frequency} = f_e = \frac{(\text{row total})(\text{column total})}{n}$$

2. A chi-square statistic is computed to measure the amount of discrepancy between the ideal sample (expected frequencies from H_0) and the actual sample data (the observed frequencies = f_o). A large discrepancy results in a large value for chi-square and indicates that the data do not fit the null hypothesis and the hypothesis should be rejected. The calculation of chi-square is the same for all chi-square tests:

   $$\text{chi-square} = \chi^2 = \sum \frac{(f_o - f_e)^2}{f_e}$$

The fact that chi-square tests do not require scores from an interval or ratio scale makes these tests a valuable alternative to the t tests, ANOVA, or correlation, because they can be used with data measured on a nominal or an ordinal scale.

Measuring Effect Size for the Chi-Square Test for Independence

When both variables in the chi-square test for independence consist of exactly two categories (the data form a 2x2 matrix), it is possible to re-code the categories as 0 and 1 for each variable and then compute a correlation known as a **phi-coefficient** that measures the strength of the relationship. (The phi-coefficient was introduced in Chapter 16.) The value of the phi-coefficient, or the squared value which is equivalent to an r^2, is used to measure the effect size. When there are more than two categories for one (or both) of the variables, then you can measure effect size using a modified version of the phi-coefficient known as Cramér's V. The value of V is evaluated much the same as a correlation.

LEARNING OBJECTIVES

1. Recognize the experimental situations where a chi-square test is appropriate.

2. Be able to conduct a chi-square test for goodness of fit to evaluate a hypothesis about the shape of a population frequency distribution.

3. Be able to conduct a chi-square test for independence to evaluate a hypothesis about the relationship between two variables.

4. Be able to measure effect size for the test for independence using either the phi-coefficient or Cramer's V.

NEW TERMS AND CONCEPTS

The following terms were introduced in this chapter. You should be able to define or describe each term and, where appropriate, describe how each term is related to other terms in the list.

Parametric statistical tests	A test evaluating hypotheses about population parameters and making assumptions about parameters. Also, a test requiring numerical scores.
Non-parametric statistical tests	A test that does not test hypotheses about parameters or make assumptions about parameters. The data usually consist of frequencies.
Expected frequencies	Hypothetical, ideal frequencies that are predicted from the null hypothesis.
Observed frequencies	The actual frequencies that are found in the sample data.
Chi-square statistic	A test statistic that evaluates the discrepancy between a set of observed frequencies and a set of expected frequencies.

Chi-square distribution   The theoretical distribution of chi-square values that would be obtained if the null hypothesis was true.

Chi-square test for goodness of fit   A test that uses the proportions found in sample data to test a hypothesis about the corresponding proportions in the general population.

Chi-square test for independence   A test that uses the frequencies found in sample data to test a hypothesis about the relationship between two variables in the population.

Phi-coefficient   A correlational measure of relationship when both variables consist of exactly two categories. A measure of effect size for the test for independence.

Cramér=s V   A modification of the phi-coefficient to be used when one or both variables consist of more than two categories

NEW FORMULAS

$$\chi^2 = \Sigma \frac{(f_o - f_e)^2}{f_e}$$

$$f_e = pn \quad \text{(test for goodness of fit)}$$

$$f_e = \frac{(\text{Row Total})(\text{Column Total})}{n} \quad \text{(test for independence)}$$

$$\text{phi} = \varphi = \frac{\chi^2}{n} \qquad \text{Cramer's V} = \frac{\chi^2}{n(df^*)}$$

where df^* is the smaller of $C - 1$ and $R - 1$.

STEP BY STEP

<u>The chi-square test for independence</u>: The chi-square test for independence uses frequency data to test a hypothesis about the relationship between two variables. The null hypothesis states that the two variables are independent (no relationship). Rejecting H_0 indicates that the data provide convincing evidence of a consistent relationship between the two variables. The following example will be used to demonstrate this chi-square test.

A psychologist would like to examine preferences for the different seasons of the year and how these preferences are related to gender. A sample of 200 people is obtained and the individuals are classified by sex and preference. The psychologist would like to know if there is a consistent relationship between sex and preference. The frequency data are as follows:

Favorite Season

	Summer	Fall	Winter	Spring	
Males	28	32	15	45	80
Females	32	8	5	35	120
	60	40	20	80	

Step 1: State the hypotheses and select an alpha level. The null hypothesis says that the two variables are independent.

H_0: Preference is independent of sex. One version of the null hypothesis states that there is no relationship between preference and sex. The second version of H_0 states that there is no difference between the distribution of preferences for males and the distribution of preferences for females (both distributions have the same proportions).

The alternative hypothesis simply says that the two variables are not independent.

H_1: Preference is related to sex, or the distribution of preferences is different for males and females.

We will use $\alpha = .05$

Step 2: Locate the critical region. The degrees of freedom for the chi-square test for independence are
   $df = (C - 1)(R - 1)$
For this example, $df = 3(1) = 3$. Sketch the distribution and locate the extreme 5%. The critical boundary is 7.81.

Step 3: Compute the test statistic. The major concern for this chi-square test is determining the expected frequencies. We begin with a blank matrix showing only the row and column totals from the data.

Favorite Season

	Summer	Fall	Winter	Spring	
Males					80
Females					120
	60	40	20	80	

The expected frequencies can be computed using the equation, or they can be determined directly from the null hypothesis. In this example, H_0 says that the distribution of preferences is the same for both genders. Therefore, we must determine the "distribution of preferences."

For the total sample of n = 200 the data show:
  60/200 = 30% prefer Summer
  40/200 = 20% prefer Fall
  20/200 = 10% prefer Winter
  80/200 = 40% prefer Spring

Next, we apply this distribution to each gender group.

There are 120 males. Using the proportions from the overall distribution, we would expect:
  30% of 120 = 36 males prefer summer
  20% of 120 = 24 males prefer fall
  10% of 120 = 12 males prefer winter
  40% of 120 = 48 males prefer spring

For the group of 80 females we would expect:
  30% of 80 = 24 females prefer summer
  20% of 80 = 16 females prefer fall
  10% of 80 = 8 females prefer winter
  40% of 80 = 32 females prefer spring

Place these values in a matrix of expected frequencies as follows:

Favorite Season

	Summer	Fall	Winter	Spring	
Males	36	24	12	48	80
Females	24	16	8	32	120
	60	40	20	80	

Now you are ready to compute the chi-square statistic.
a) For each cell in the matrix, find the difference between the expected and the observed frequency.

b) Square the difference.
c) Divide the squared difference by the expected frequency.
d) Sum the resulting values for each category

f_o	f_e	$(f_o - f_e)$	$(f_o - f_e)^2$	$(f_o - f_e)^2/f_e$
28	36	−8	64	1.78
32	24	8	64	2.67
15	12	3	9	0.75
45	48	−3	9	0.19
32	24	8	64	2.67
8	16	−8	64	4.00
5	8	−3	9	1.13
35	32	3	9	0.28
				13.47 = χ^2

Step 4: Make decision. The chi-square value is in the critical region. Therefore, we reject H_0 and conclude hat there is a significant relation between sex and preferred season.

To describe the nature of the relationship, you can compare the data with the expected frequencies. From this comparison it should be clear that more men prefer fall and more women prefer summer than would be expected by chance.

HINTS AND CAUTIONS

1. When computing expected frequencies for either chi-square test, it is wise to check your arithmetic by being certain that $\Sigma f_e = \Sigma f_o = n$. In the test for independence, the expected frequencies in any row or column should sum to the same total as the corresponding row or column in the observed frequencies.

2. Whenever a chi-square test has df = 1, the difference (absolute value) between f_o and f_e will be the same for every category. This is true for either the goodness of fit test with 2 categories, or the test of independence with 4 categories. Knowing this fact can help you check the calculation of expected frequencies and it can simplify the calculation of the chi-square statistic.

SELF TEST

True/False Questions

1. One characteristic of nonparametric tests is that they make few, if any, assumptions about the populations being investigated.

2. The chi-square test for goodness of fit compares two separate samples to determine whether the two distributions have the same shape (same proportions).

3. The data for a chi-square test are called observed frequencies.

4. The observed frequencies for a chi-square test are always whole numbers (no fractions or decimals).

5. The expected frequencies for a chi-square test are always whole numbers (no fractions or decimals).

6. A large value for the chi-square statistic is an indication that the data fit the hypothesis.

7. The chi-square distribution tends to be symmetrical with a mean of $\mu = 0$.

8. The degrees of freedom for a chi-square test are not related to the size of the sample.

9. The data for a chi-square test for independence can be viewed either as representing one sample with two measurements for each participant or as two (or more) separate samples.

10. The phi-coefficient can be used to measure effect size for a chi-square test for independence provided there are exactly two categories for each of the two variables.

Multiple-Choice Questions

1. Which of the following is a characteristic of nonparametric tests?
    a. they require a numerical score for each individual
    b. they require assumptions about the population distribution(s)
    c. they evaluate hypotheses about population means or variances
    d. None of the other choices is a characteristic of a nonparametric test.

2. Which of the following best describes the observed frequencies for a chi-square test?
    a. They are positive and are always whole numbers.
    b. They are positive and can contain fractions or decimal values.
    c. They can be either positive or negative values, but are always whole numbers.
    d. They can be either positive or negative values and can contain fractions or decimals.

3. Which of the following best describes the expected frequencies for a chi-square test?
    a. They are positive and are always whole numbers.
    b. They are positive and can contain fractions or decimal values.
    c. They can be either positive or negative values, but are always whole numbers.
    d. They can be either positive or negative values and can contain fractions or decimals.

4. Which of the following best describes the possible values for a chi-square statistic?
    a. Chi-square is always a positive whole numbers.
    b. Chi-square is always positive but can contain fractions or decimal values.
    c. Chi-square can be either positive or negative, but always is a whole number.
    d. Chi-square can be either positive or negative and can contain fractions or decimals.

5. Which of the following describes the typical chi-square distribution?
    a. It is symmetrical, centered at a value of zero.
    b. It is symmetrical, centered at a value determined by the degrees of freedom.
    c. It is positively skewed, with no values less than zero.
    d. It is negatively skewed, with no values less than zero.

6. How does the difference between f_e and f_o influence the outcome of a chi-square test?
    a. The larger the difference, the larger the value of chi-square and the greater the likelihood of rejecting the null hypothesis.
    b. The larger the difference, the larger the value of chi-square and the lower the likelihood of rejecting the null hypothesis.
    c. The larger the difference, the smaller the value of chi-square and the greater the likelihood of rejecting the null hypothesis.
    d. The larger the difference, the smaller the value of chi-square and the lower the likelihood of rejecting the null hypothesis.

7. A researcher is using a chi-square test to determine whether people have any preferences among three brands of televisions. The null hypothesis for this test would state that _____.
    a. there are real preferences in the population
    b. one-third of the sample will prefer each brand
    c. one-third of the population prefers each brand
    d. in the population, one brand is preferred over the other two

8. A chi-square test for goodness of fit has df = 2. How many categories were used to classify the individuals in the sample?
    a. 2
    b. 3
    c. 4
    d. cannot be determined without additional information

9. A sample of n = 100 people is classified into four categories. If the results are evaluated with a chi-square test for goodness of fit, what is the df value for the chi-square statistic?
    a. 3
    b. 4
    c. 99
    d. 100

10. For a fixed α level, how is the critical value for chi-square related to the size of the sample?
    a. As the sample size increases, the critical value also increases.
    b. As the sample size increases, the critical value decreases.
    c. The critical value of chi-square is not related to the sample size.

11. For a fixed level of significance, the critical value for chi-square will _____.
    a. increase when df increases.
    b. decrease when df increases.
    c. increase when n and df both increase.
    d. The critical value is not related to either n or df.

12. A sample of 100 people is classified by gender (male/female) and by whether or not they are registered voters. The sample consists of 80 females and 20 males, and has a total of 60 registered voters. If these data were used for a chi-square test for independence, the expected frequencies should have a total of ____ males.
    a. 12
    b. 20
    c. 40
    d. 48

13. A sample of 100 people is classified by gender (male/female) and by whether or not they are registered voters. The sample consists of 80 females and 20 males, and has a total of 60 registered voters. If these data were used for a chi-square test for independence, what is the expected frequency for males who are registered voters?
    a. 12
    b. 20
    c. 40
    d. cannot determine without additional information

14. A chi-square test for independence is being used to evaluate the relationship between two variables, one of which is classified into 3 categories and the second of which is classified into 4 categories. The chi-square statistic for this test would have df equal to _____.
    a. 6
    b. 7
    c. 10
    d. 11

15. When the data form a 2x2 matrix, the phi-coefficient is used to measure effect size for the chi-square test for independence. If other factors are held constant, how does sample size influence the values for phi and chi-square?
    a. A larger sample increase both phi and chi-square.
    b. A larger sample increases phi but has no effect on chi-square.
    c. A larger sample increases chi-square but has no effect on phi.
    d. Sample size does not influence either phi or chi-square.

Other Questions

1. A researcher is examining preferences among four new flavors of ice cream. A sample of n = 80 people is obtained. Each person tastes all four flavors and then picks his/her favorite. The distribution of preferences is as follows.

Ice Cream Flavor

A	B	C	D
12	18	28	22

Do these data indicate any significant preferences among the four flavors? Test at the .05 level of significance.

2. A researcher is testing four new flavors of bubble gum using a sample that consists of 50 men, 200 women, and 250 children. Each individual selects his/her favorite flavor. In the total sample of 500 people, 100 selected Flavor A, 200 chose B, 150 picked C, and only 50 preferred D.
   a. If these data are used for a chi-square test for independence, state the null hypothesis.
   b. Use the following matrix to fill in the expected frequencies for the chi-square test for independence.

Flavor

	A	B	C	D	
Men					50
Women					200
Children					250
	100	200	150	50	

3. A researcher is interested in the relationship between birth order and personality. A sample of n = 100 people is obtained, all of whom grew up in families as one of three children. Each person is given a personality test and the researcher also records the person's birth order position (1st born, 2nd, or 3rd). The frequencies from this study are shown in the following table. On the basis of these data can the researcher conclude that there is a significant relation between birth order and personality? Test at the .05 level of significance.

Birth Position

	1st	2nd	3rd
Outgoing	13	31	16
Reserved	17	19	4

ANSWERS TO SELF TEST

True/False Answers

1. True
2. False. The test for goodness of fit uses one sample to test a hypothesis about one population.
3. True
4. True
5. False. Expected frequencies are computed and can be fractions or decimals.
6. False. A large value for chi-square indicates that the data do not fit the hypothesis and H_0 should be rejected.
7. False. The chi-square distribution is positively skewed with all value greater than or equal to zero.
8. True
9. True
10. True

Multiple-Choice Answers

1. d Nonparametric tests do not involve parameters and make few if any assumptions about populations.
2. a The observed frequencies are obtained by counting individuals in the sample.
3. b The expected frequencies are calculated values that represent an ideal sample in perfect accord with the null hypothesis.
4. b Chi square is a sum of squared values so it is always positive and the distribution is cut off at zero.
5. c Chi-square is a distribution of squared numbers that should be relatively small if the null hypothesis is true.
6. a The chi-square statistic measures the difference between the data and the hypothesis.
7. c Equal preferences implies equal proportions in the population.
8. b df = the number of categories minus one for the goodness of fit test.
9. a df is determined by the number of categories and has no relation to the sample size.
10. c df is determined by the number of categories and has no relation to the sample size.
11. c The larger the df value, the more categories that are being summed to obtain chi-square.
12. b The column totals and row totals must be the same for the observed frequencies and the expected frequencies.
13. a 60% of 20 males = 12
14. a $(3 - 1)(4 - 1) = 6$
15. c As with all tests, sample size influences the hypothesis test (chi-square) but has little or no effect on measures of effect size (phi).

Other Answers

1. The null hypothesis states that there are no preferences among the four flavors and that 1/4th of the population should prefer each flavor. The expected frequency are 20 for each flavor. With df = 3, the critical value is 7.81. For these data, the chi-square statistic is 6.80. The decision is to fail to reject the null hypothesis and conclude that the data do not provide sufficient evidence to indicate significant preferences among the four ice cream flavors.

2. a. One version of the null hypothesis states that gum preference is independent of the age/gender classifications. The second version of H_0 states that the distribution of flavor preferences is the same (same proportions) for men, women, and children.

   b.

	Flavor A	Flavor B	Flavor C	Flavor D	
Men	10	20	15	5	50
Women	40	80	60	20	200
Children	50	100	75	25	250
	100	200	150	50	

3. The null hypothesis states that there is no relation between birth order and personality - the two variables are independent. With df = 2, the critical value for this test is 5.99. The expected frequencies are as follows:

	1st	2nd	3rd
Outgoing	18	30	12
Reserved	12	20	8

Birth Position

For these data, the chi-square statistic is 6.89. Reject H_0 and conclude that there is a significant relation between personality and birth order.

CHAPTER 19

THE BINOMIAL TEST

CHAPTER SUMMARY

The goals of Chapter 19 are:
1. To introduce the binomial test, using the normal approximation to the binomial distribution, as a hypothesis testing procedure for binomial data.
2. To introduce the sign test as a special application of the binomial test.

The Binomial Test

The binomial test provides a method for testing hypotheses about population proportions for populations consisting of binomial data. **Binomial data** exist when the measurement procedure classifies individuals into exactly two distinct categories. Traditionally, the two categories are identified as A and B, and the population proportions are identified as $p(A) = p$ and $p(B) = q$. The null hypothesis specifies the values of p and q for the population. For example, when testing whether or not a coin is balanced, the null hypothesis would state that the coin is balanced or, $p(\text{Heads}) = p = .50$, and $p(\text{Tails}) = q = .50$.

The sample data for the binomial test consist of a sample of n individuals each of whom is classified in category A or B. The sample statistic, X, is simply the number of individuals classified in category A. The logic underlying the **binomial test** is identical to the logic for the original z-score test (Chapter 8) or the t-statistic hypothesis tests (Chapters 9, 10, and 11). The test statistic compares the sample data with the hypothesized value for the population. If the data are consistent with the hypothesis, we conclude that the hypothesis is reasonable. However, if there is a large discrepancy between the data and the hypothesis, we reject the hypothesis.

When the values of pn and qn are both greater than or equal to 10, the binomial distribution is approximately normal with a mean of $\mu = pn$ and a standard deviation of $\sigma = \sqrt{npq}$ (see Chapter 6). In this case, the binomial test can be conducted by transforming the X value from the sample into a z-score and then using the unit normal table to determine critical values. If the z-score is only slightly into the critical region, you should check both real limits for X to ensure that the entire score is beyond the critical boundary.

The Sign Test

The **sign test** is a special application of the binomial test used to evaluate the results from a repeated-measures research design comparing two treatment conditions. The difference score for each individual is classified as either an increase (+) or a decrease (−) and the binomial test evaluates a null hypothesis stating that increases and decreases are equally likely: $p(+) = p(-) = 1/2$.

LEARNING OBJECTIVES

1. Recognize binomial data and identify situations where a binomial test is appropriate.

2. Understand the criteria that must be satisfied before the binomial distribution is normal and the z-score statistic can be used.

3. Be able to perform all of the necessary computations for the binomial test.

4. Recognize when a sign test is appropriate and be able to perform the necessary computations.

NEW TERMS AND CONCEPTS

The following terms were introduced in this chapter. Define or describe each term and, where appropriate, describe how each term is related to other terms in the list.

Binomial data — Scores obtained from a scale of measurement that consists of exactly two categories.

Binomial distribution — The distribution of probabilities for all the possible outcomes for a set of binomial measurements.

Binomial test — A test to determine whether or not a set of binomial sample data deviate significantly from the population binomial proportions specified in a null hypothesis.

Sign test — A special application of the binomial test used with data from a repeated-measures study comparing two treatments. The test determines whether the increases and decreases in the difference scores are significantly different from what would be expected just by chance.

NEW FORMULAS

$$z = \frac{X - pn}{\sqrt{npq}} \quad \text{or} \quad z = \frac{X/n - p}{\sqrt{pq/n}}$$

THE BINOMIAL TEST

STEP BY STEP

The Binomial Hypothesis Test: The binomial test presented in this chapter is used to test a hypothesis about unknown population proportions using binomial data from a single sample. Interpretation of the z-score statistic requires that the values of pn and qn are both at least 10 so the binomial distribution is approximately normal. The binomial test uses the same four-step procedure that we use for all hypothesis tests. We will use the following example to demonstrate the binomial test.

A survey of last year's graduating college seniors showed that 40% intended to continue their education with some form of graduate training. In a sample of n = 54 seniors from this year's class only X = 10 indicated that they planned to go on for graduate training. On the basis of this sample can you conclude that this year's class has significantly different plans than last year's class? Test at the .05 level of significance.

Step 1: State Hypotheses and select an alpha level. The null hypothesis states that the proportions for this year's class are not different from last year's class.
H_0: p = p(continue education) = 0.40 (no change)
H_1: p ≠ 0.40 (this year's proportion is different)
For this example we are using α = .05

Step 2: Locate the critical region. For this example pn = 0.40(54) = 21.6 and qn = 0.60(54) = 32.4. Both values are greater than 10, so the normal approximation is appropriate. For a two-tailed test with α = .05 the critical z-score values are z = ∀1.96.

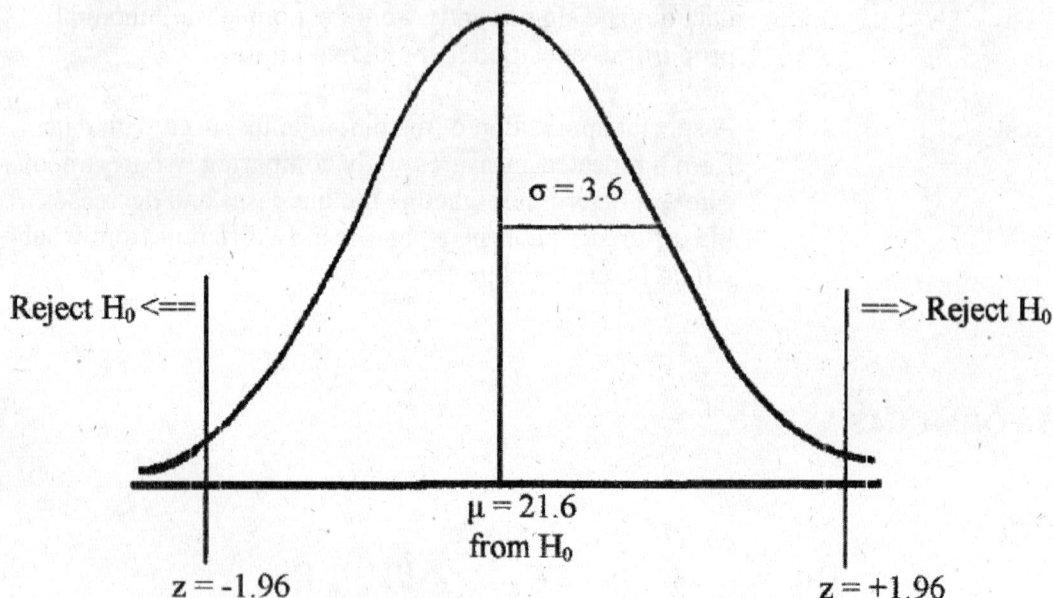

Step 3: Compute the test statistic: For this sample we have X = 10. Using n = 54 and the hypothesized values of p = 0.40 and q = 0.60 from H_0, the z-score statistic is

$$z = \frac{X - pn}{\sqrt{npq}} = \frac{10 - (0.40)(54)}{\sqrt{54(0.4)(0.6)}} = \frac{10 - 21.6}{3.6} = -3.222$$

Step 4: Make decision. The z-score we obtained is well into the critical region. This is a very unlikely value to obtain by chance ($p < .05$). Therefore, we reject H_0 and conclude that these data provide sufficient evidence to demonstrate that this year's graduating class has significantly different proportions than last year's class.

HINTS AND CAUTIONS

1. Remember, the values of pn and qn must *both* be greater than or equal to 10 before the z-score normal approximation can be used.

2. When you are discarding zero differences with the sign test, remember to reduce the value of n appropriately.

3. Remember that the normal distribution is continuous and only approximates the discrete binomial distribution. If the z-value obtained in a binomial test is only slightly into the critical region, check both real limits for X to make sure that the entire interval corresponding to the X value is in the critical region

SELF TEST

True/False Questions

1. The null hypothesis for a binomial test specifies the values of p and q for the sample.

2. In a binomial test, the value of X can be any whole number from 0 to n.

3. For a binomial test, X is the number of individuals in one of the two categories.

4. For a binomial test, if $p = 0.20$ then q must equal 0.80.

5. The binomial distribution for $p = 0.80$ and $n = 30$ satisfies the criterion for using the normal approximation.

6. For $p = 0.20$, the binomial distribution for a sample of $n = 80$ will have a mean of $\mu = \sqrt{16} = 4$.

7. For $p = 0.20$, the binomial distribution for a sample of $n = 100$ will have a standard deviation of $\sigma = \sqrt{16} = 4$.

8. The data from a binomial test can also be evaluated using a chi-square test for goodness of fit.

9. The sign test is a special application of the binomial test used with data from an independent-measures study.

10. For a sign test, the null hypothesis always specifies $p = q = 1/2$.

Multiple-Choice Questions

1. If a binomial distribution has $p = 0.60$, then what is the value for q?
    a. 0.36
    b. 0.40
    c. 0.60
    d. 1.60

2. For a binomial test, if $pn = 30$ and $qn = 90$, what is the value of n?
    a. 60
    b. 100
    c. 120
    d. 150

3. For a binomial test, if $pn = 30$ and $qn = 90$, what is the value of p?
    a. $p = 0.25$
    b. $p = 0.30$
    c. $p = 0.60$
    d. $p = 0.67$

4. For a sample of $n = 200$, what is the mean of the binomial distribution with $p = .20$ and $q = .80$?
    a. 20
    b. 40
    c. 80
    d. 160

5. For a sample of $n = 100$, what is the standard deviation of the binomial distribution with $p = .20$ and $q = .80$?
    a. 2
    b. 4
    c. 8
    d. 16

6. If p = .20, how large a sample is needed to justify using the normal approximation to the binomial distribution.
   a. 20
   b. 40
   c. 50
   d. 100

7. The probability of correctly predicting the outcome of a coin toss is p = ½. What is the mean of the binomial distribution for the number of correct predictions in a series of 36 tosses?
   a. 3
   b. 9
   c. 18
   d. √18

8. On a multiple-choice exam for which each question has four possible answers, the probability of guessing the correct answer is p = 1/4. What is the standard deviation of the normal approximation to the binomial distribution for the number of correct guesses for a series of 48 questions?
   a. 3
   b. 9
   c. 12
   d. 24

9. For a binomial distribution with p = .25 and n = 48, what is the z-score corresponding to X = 15?
   a. z = 3/3 = 1.00
   b. z = 3/9 = 0.33
   c. z = –9/3 = –3.00
   d. z = –9/9 = –1.00

10. For a binomial distribution with p = 1/5 and n = 100, what is the z-score corresponding to X = 26?
    a. 0.30
    b. 0.60
    c. 1.00
    d. 1.50

11. A binomial test produces z = 4.00. If a chi-square test is used to evaluate the same data, what value will be obtained for chi-square?
    a. $\chi^2 = \sqrt{2.00}$
    b. $\chi^2 = 2.00$
    c. $\chi^2 = 4.00$
    d. $\chi^2 = 16.00$

12. For a sign test with n = 36, what are the mean and standard deviation for the binomial distribution.
   a. μ = 18 and σ = 9
   b. μ = 18 and σ = 3
   c. μ = 18 and σ = √3
   d. Cannot answer without additional information.

13. For a sign test, what is the consequence of discarding individuals with zero difference scores instead of dividing them equally between the increases and decreases?
   a. You increase the likelihood of rejecting the null hypothesis.
   b. You decrease the likelihood of rejecting the null hypothesis.
   c. You increase the likelihood of a Type II error.
   d. There is no difference between the two methods for dealing with zero differences.

14. In the sign test, zero differences _____.
   a. provide support for the null hypothesis
   b. provide evidence to reject the null hypothesis
   c. are not relevant to the null hypothesis and should be discarded
   d. indicate that assumptions have been violated and the test should not be done

15. For a binomial distribution with p = .10 and n = 100, what is the z-score corresponding to X = 16?
   a. 6
   b. 2
   c. 1.6
   d. 3

Other Questions

1. In a blind taste test of two leading brands of herbal tea, 25 out of 36 people preferred brand A. Do these data indicate a significant preference for brand A over brand B? Test at the .05 level of significance.

2. According to most estimates, 10% of the people in the general population are left-handed and 90% are right-handed. A neurological theory of handedness and cerebral dominance predicts that there should be a disproportionate number of left-handers among artists. For a sample of n = 150 artists, a psychologist finds that 24 are left-handed. Is there a higher proportion of left-handers among artist than there is for the general population? Test with alpha set at .05.

3. An instructor is curious if students cut more or fewer classes in the second half of the semester than in the first half. For a sample of n = 37 students, the instructor records the number of classes each student missed up to midterm and then how many each missed from midterm to the end of the semester. Twenty two students showed an increase in

cuts from the first half to the second half of the semester, 14 showed a decrease, and 1 showed no change. Are significantly more classes cut in one part of the semester than the other? Test with α = .05.

4. A test for ESP requires a subject to predict the suit of a playing card randomly selected from a complete deck. If the test involves a series of n = 40 predictions, how many would a subject need to predict correctly to do significantly better than chance. Assume a two-tailed test with α = .05.

5. A company embarks on a new program to boost morale and improve employee productivity. Records for a sample of n = 64 employees indicate that 39 showed improved productivity after the program was initiated, and 46 reported improved morale.
   a. Do these data indicate that the program had a significant effect on productivity? Test at the .01 level of significance.
   b. Do the data indicate a significant effect on morale? Again, test at the .01 level of significance.

ANSWERS TO SELF TEST

True/False Answers

1. False. The null hypothesis identifies p and q for the population.
2. True
3. True
4. True
5. False. The binomial distribution is normal if both pn and qn are greater than or equal to 10. In this case, qn = .2(30) = 6.
6. False. $\mu = pn = 16$
7. True
8. True
9. False The sign test is used with repeated-measures designs.
10. True

Multiple-Choice Answers

1. b p + q = 1.00
2. c pn + qn = n
3. a pn + qn = n = 120. pn = 30.
4. b The mean for the binomial distribution is pn = 40.
5. b The standard deviation is the square root of (npq).
6. c pn and qn must both be at least 10.
7. c For the normal approximation, $\mu = pn$.
8. a For the normal approximation, σ equals the square root of (npq)

9. a With p = 1/4 and n = 48, the distribution has $\mu = 12$ and $\sigma = 3$. X = 15 corresponds to z = 3/3 = 1.00.
10. d With p = 1/5 and n = 100, the distribution has $\mu = 20$ and $\sigma = 4$. X = 26 corresponds to z = 6/4 = 1.50.
11. d $\chi^2 = z^2$
12. b With p = q = ½, the mean is 18 and the standard deviation is 3.
13. a Discarding zero differences is discarding individuals who support the null hypothesis.
14. a The null hypothesis says that there is no difference between the two treatments.
15. b With p = 0.10 and n = 100, the distribution has $\mu = 10$ and $\sigma = 3$. X = 16 corresponds to z = 6/3 = 2.00.

Other Answers

1. H_0: p = q = 1/2 (no preference). With $\mu = 18$ and $\sigma = 3$, X = 25 out of 36 corresponds to z = 2.33. Reject H_0 and conclude that there is a significant preference for brand A.

2. H_0: p = 0.1 left-handed and q = 0.9 right-handed. X = 24 out of 150 corresponds to z = 2.45. Reject H_0 and conclude that there is a significant difference in the proportion of left-handers among artists.

3. H_0: p = q = 1/2 (increases and decreases are equally likely). Discarding the student who showed no difference leaves n = 36. With $\mu = 18$ and $\sigma = 3$, X = 22 out of 36 corresponds to z = 1.33. Fail to reject H_0 and conclude that there is not a significant difference between the two halves of the semester.

4. With $\alpha = .05$, the subject must have a z-score greater than 1.96 to be significantly better than chance. By chance, you would expect p = 1/4 correct and q = 3/4 incorrect predictions. With n = 40, the standard error is $\sqrt{npq}$ = $\sqrt{40(/1/4)(3/4)}$ = 2.74. The subject must be above chance by at least 2.74(1.96) = 5.37 to be in the critical region. With a chance level of (1/4)(40) = 10, the subject must get at least X = 15.37 (X = 16) to be significantly better than chance.

5. a. H_0: p = q = 1/2 (increases and decreases are equally likely). The sample value, X = 39, corresponds to z = 1.75. Fail to reject H_0 and conclude that these data are not sufficient to conclude that there has been a significant change in productivity.
   b. H_0: p = q = 1/2 (increases and decreases are equally likely). The sample value, X = 46, corresponds to z = 3.50. Reject H_0 and conclude that these data are sufficient to conclude that there has been a significant change in employee morale.

CHAPTER 20

STATISTICAL TESTS FOR ORDINAL DATA

CHAPTER SUMMARY

The goals of Chapter 20 are:
1. To introduce the concept that ordinal data (ranks) can be used to evaluate treatments and differences between treatments, but they require special techniques.
2. To demonstrate the Mann-Whitney U-test as a technique for evaluating the difference between two separate samples.
3. To demonstrate the Wilcoxon Test as a technique for evaluating the difference between two treatments using data from a repeated-measures design.
4. To demonstrate the Kruskal-Wallis test as a technique for evaluating the differences between three or more separate samples.
5. To demonstrate the Friedman test as a technique for evaluating the differences between three or more treatments using data from a repeated-measures design.

Tests for Ordinal Data
   Chapter 20 introduces four statistical techniques that have been developed specifically for use with **ordinal data**; that is, data where the measurement procedure simply arranges the subjects into a rank-ordered sequence. The statistical methods presented in this chapter can be used when the original data consist of ordinal measurements (ranks), or when the original data come from an interval or ratio scale but are converted to ranks because they do not satisfy the assumptions of a standard parametric test such as the t statistic. Four statistical methods are introduced:
1. The Mann-Whitney test evaluates the difference between two treatments or two populations using data from an independent-measures design; that is, two separate samples.
2. The Wilcoxon test evaluates the difference between two treatment conditions using data from a repeated-measures design; that is, the same sample is tested/measured in both treatment conditions.
3. The Kruskal-Wallis test evaluates the differences between three or more treatments (or populations) using a separate sample for each treatment condition.
4. The Friedman test evaluates the differences between three or more treatments for studies using the same group of participants in all treatments (a repeated-measures study).

The Mann-Whitney U Test
   The **Mann-Whitney test** can be viewed as an alternative to the independent-measures t test (Chapter 10). The test uses the data from two separate samples to test for

a significant difference between two treatments or two populations. However, the Mann-Whitney test only requires that you are able to rank order the individual scores; there is no need to compute means or variances. The null hypothesis for the Mann-Whitney test simply states that there is no systematic or consistent difference between the two treatments (populations) being compared. The calculation of the Mann-Whitney U statistic requires:
1. Combine the two samples and rank order the individuals in the combined group.
2. Once the complete set is rank ordered, you can compute the Mann-Whitney U by either
   a. Find the sum of the ranks for each of the original samples, and use the formula to compute a U statistic for each sample.
   b. Compute a score for each sample by viewing the ranked data (1st, 2nd, etc.) as if they were finishing positions in a race. Each subject in sample A receives one point for every individual from sample B that he/she beats. The total number of points is the U statistic for sample A. In the same way, a U statistic is computed for sample B.
3. The Mann-Whitney U is the smaller of the two U statistics computed for the samples.

If there is a consistent difference between the two treatments, the two samples should be distributed at opposite ends of the rank ordering. In this case, the final U value should be small. At the extreme, when there is no overlap between the two samples, you will obtain $U = 0$. Thus, a small value for the Mann-Whitney U indicates a difference between the treatments. To determine whether the obtained U value is sufficiently small to be significant, you must consult the Mann-Whitney table. For large samples, the obtained U statistic can be converted to a z-score and the critical region can be determined using the unit normal table.

The Wilcoxon T test

The **Wilcoxon test** can be viewed as an alternative to the repeated-measures t test (Chapter 11). The test uses the data from one sample where each individual has been observed in two different treatment conditions to test for a significant difference between the two treatments. However, the Wilcoxon test only requires that you are able to rank order the difference scores; there is no need to measure how much difference exists for each subject, or to compute a mean or variance for the difference scores. The null hypothesis for the Wilcoxon test simply states that there is no systematic or consistent difference between the two treatments being compared. The calculation of the Wilcoxon T statistic requires:
1. Observe the difference between treatment 1 and treatment 2 for each subject.
2. Rank order the absolute size of the differences without regard to sign (increases are positive and decreases are negative).
3. Find the sum of the ranks for the positive differences and the sum of the ranks for the negative differences.
4. The Wilcoxon T is the smaller of the two sums.

If there is a consistent difference between the two treatments, the difference scores should be consistently positive (or consistently negative). At the extreme, all the differences will be in the same direction and one of the two sums will be zero. (If there are no negative differences then $\Sigma Ranks = 0$ for the negative differences.) Thus, a small value for T indicates a difference between treatments. To determine whether the obtained T value is sufficiently small to be significant, you must consult the Wilcoxon table. For large samples, the obtained T statistic can be converted to a z-score and the critical region can be determined using the unit normal table.

The Kruskal-Wallis Test

The **Kruskal-Wallis test** can be viewed as an alternative to a single-factor, independent-measures analysis of variance. The test uses data from three or more separate samples to evaluate differences among three or more treatment conditions. The test requires that you are able to rank order the individuals but does not require numerical scores. The null hypothesis for the Kruskal-Wallis test simply states that there are no systematic or consistent differences among the treatments being compared. The calculation of the Kruskal-Wallis statistic requires:

1. Combine the individuals from all the separate samples and rank order the entire group.
2. Regroup the individuals into the original samples and compute the sum of the ranks (T) for each sample.
3. A formula is used to compute the Kruskal-Wallis statistic which is distributed as a chi-square statistic with degrees of freedom equal to the number of samples minus one. The obtained value must be greater than the critical value for chi-square to reject H_0 and conclude that there are significant differences among the treatments.

The Friedman test

The **Friedman test** can be viewed as an alternative to a single-factor, repeated-measures analysis of variance. The test uses data from one sample to evaluate differences among three or more treatment conditions. The test requires that you are able to rank order the individuals across treatments but does not require numerical scores. The null hypothesis for the Friedman test simply states that there are no systematic or consistent differences among the treatments being compared. The calculation of the Friedman statistic requires:

1. Each individual (or the individual's scores) must be ranked across the treatment conditions.
2. The sum of the ranks (R) is computed for each treatment.
3. A formula is used to compute the Friedman test statistic which is distributed as chi-square with degrees of freedom equal to the number of treatments minus one. The obtained value must be greater than the critical value for chi-square to reject H_0 and conclude that there are significant differences among the treatments.

LEARNING OBJECTIVES

1. Know when the Mann-Whitney U test, the Wilcoxon Signed-Ranks test, the Kruskal-Wallis test, and the Friedman test are appropriate.

2. Be able to compute and evaluate the Mann-Whitney U, the Wilcoxon T, the Kruskal-Wallis H (chi-square) and the Friedman chi-square.

NEW TERMS AND CONCEPTS

The following terms were introduced in this chapter. You should be able to define or describe each term and, where appropriate, describe how each term is related to other terms in the list.

Ordinal data	Scores from a scale of measurement that identifies direction but not distance between categories.
Mann-Whitney U	A test for a significant difference between two treatments using data from two separate samples measured on an ordinal scale.
Normal approximation to the Mann-Whitney U	With large samples, the Mann-Whitney U statistic tends to form a normal distribution with each value of U corresponding to a specific z-score.
Wilcoxon T	A test for a significant difference between two treatments using data from one sample measured in both treatments. The difference scores are measured on an ordinal scale.
Normal approximation to the Wilcoxon T	With large samples, the Wilcoxon T statistic tends to form a normal distribution with each value of T corresponding to a specific z-score.
Kruskal-Wallis test	A test for significant difference among three or more treatments using data from three or more separate samples measured on an ordinal scale
Friedman test	A test for significant differences among three or more treatments using data from a repeated-measures design (the same group of participants in all treatments).

NEW FORMULAS

$$U_A = n_A n_B + \frac{n_A(n_A + 1)}{2} - \Sigma R_A$$

$$U_B = n_A n_B + \frac{n_B(n_B + 1)}{2} - \Sigma R_B$$

$$U_A + U_B = n_A n_B$$

$$z = \frac{U - \frac{n_A n_B}{2}}{\sqrt{\frac{n_A n_B(n_A + n_B + 1)}{12}}} \quad \text{(for Mann-Whitney)}$$

$$z = \frac{T - \frac{n(n + 1)}{4}}{\sqrt{\frac{n(n + 1)(2n + 1)}{24}}} \quad \text{(for Wilcoxon)}$$

$$H \text{ (or } \chi^2) = \frac{12}{N(N + 1)} (\Sigma \frac{T^2}{n}) - 3(N + 1) \quad \text{(for Kruskal-Wallis)}$$

$$\chi^2 = \frac{12}{nk(k + 1)} \Sigma R^2 - 3n(k + 1) \quad \text{(for Friedman)}$$

STEP BY STEP

<u>The Mann-Whitney U test</u>: As a non-parametric alternative to the independent-measures t, the Mann-Whitney test uses the data from two separate samples to test hypotheses about the difference between two populations or two treatments. The following example will be used to demonstrate the Mann-Whitney U test.

STATISTICAL TESTS FOR ORDINAL DATA

A researcher obtains two random samples with n = 5 in one sample and n = 6 in the other. Treatment A is administered to the first sample and the second sample gets Treatment B. The resulting scores for the two samples are:
Sample A: 15, 12, 9, 19, 20
Sample B: 8, 10, 5, 14, 3, 6

Step 1: State the hypotheses and select an alpha level. The hypotheses for the Mann-Whitney U are stated in general terms and do not concern any specific population parameters.
H_0: There is no systematic difference between the two treatments.
H_1: There is a systematic difference that causes the scores from one sample to be generally higher than the scores from the other sample.
We will use $\alpha = .05$.

Step 2: Locate the critical region. For the Mann-Whitney test, a small value of U indicates a substantial difference between the two samples. To determine the critical value, you must consult the Mann-Whitney table. With n = 5 and n = 6 for the two samples, and with $\alpha = .05$, the critical value is U = 3. If the data produce U = 3 or smaller, we will conclude that there is a significant difference between the two treatments.

Step 3: Compute the test statistic. Although the Mann-Whitney U does not require much calculation, there are several steps involved in finding the U value.
  a. Combine the two samples and list all the scores in rank-order, from smallest to largest.
  b. For each individual in sample A, count how many of the scores from sample B have lower positions in the list. Sum these values for all the individuals in sample A. This is U_A.
  c. In the same way, find U_B.
  d. The Mann-Whitney U is the smaller of the two U values.

For these data:

Rank	1	2	3	4	5	6	7	8	9	10	11
Score	3	5	6	8	9	10	12	14	15	19	20
Sample	B	B	B	B	A	B	A	B	A	A	A

Points for
Sample A   2   1   0 0 0

Sample A has a U value of $U_A = 2 + 1 = 3$.

Using the same procedure, compute U for sample B. To check your calculations be sure that

$U_A + U_B = n_A n_B$

You should find that the smaller U is U = 3.

Step 4: Make decision. The U value of U = 3 is in the critical region. This is a very unlikely value to obtain by chance, so we reject H_0 and conclude that there is a systematic difference between the two treatments.

The Wilcoxon test: As a non-parametric alternative to the repeated-measures t, the Wilcoxon test uses data from a single sample measured in two treatment conditions. The difference scores for the sample are used to test a hypothesis about the difference between the treatments in the population. The following example will be used to demonstrate the Wilcoxon T test.

A researcher obtains a random sample of n = 7 individuals and tests each person in two different treatment conditions. The data for this sample are:

Subject	Treatment 1	Treatment 2	Difference
#1	8	24	+16
#2	12	10	−2
#3	15	19	+4
#4	31	52	+21
#5	26	20	−6
#6	32	40	+8
#7	19	29	+10

Step 1: State the hypotheses and select an alpha level. The hypotheses for the Wilcoxon test do not refer to any specific population parameter.

H_0: There is no systematic difference between the two treatments.

H_1: There is a consistent difference between the treatments that causes the scores in one treatment to be generally higher than the scores in the other treatment.

We will use α = .05.

Step 2: Locate the critical region. A small value for the Wilcoxon T indicates that the difference scores were consistently positive (or consistently negative) which indicates a systematic treatment difference. Thus, small values will tend to refute H_0. To locate the critical value, consult the Wilcoxon table with n = 7 and α = .05. For these data a Wilcoxon T of 2 or smaller is needed to reject H_0.

Step 3: Compute the test statistic. The calculation of the Wilcoxon T is very simple, but requires several stages.
    a. Ignoring the signs (+ or −), rank the difference scores from smallest to largest.
    b. Compute the sum of the ranks for the positive differences and the sum for the negative differences.
    c. The Wilcoxon T is the smaller of the two sums.

For these data:

Difference	Rank		
(+) 16	6		
(−) 2	1		
(+) 4	2	$\Sigma R_{positive} = 24$	
(+) 21	7		The Wilcoxon T = 4
(−) 6	3	$\Sigma R_{negative} = 4$	
(+) 8	4		
(+) 10	5		

Step 4: Make decision. The obtained T value is not in the critical region. These data are not significantly different from chance. Therefore, we fail to reject H_0 and conclude that there is not sufficient evidence to conclude that there is a systematic difference between the two treatments.

Kruskal-Wallis Test: As a non-parametric alternative to analysis of variance, the Kruskal-Wallis test uses data from three or more separate samples representing three or more treatment conditions. The individuals in all the samples are combined into a single group and rank ordered. The following example shows three treatment conditions with three samples, each with n = 5 individuals. All N = 15 individuals have been ranked and the ranks are shown for each treatment condition.

Treatments		
I	II	III
3	1	9
4	2	11
7	5	13
10	6	14
12	8	15
T = 36	T = 22	T = 62

Step 1: State the hypotheses and select an alpha level. The hypotheses for the Kruskal-Wallis test do not refer to any specific population parameters.
  H_0: There are no systematic differences among the three treatments.
  H_1: There are consistent differences between the treatments that causes the scores in at least one treatment to be generally higher or lower than the scores in another treatment.
We will use $\alpha = .05$.

Step 2: Locate the critical region. The test statistic for Kruskal-Wallis is distributed as chi-square with degrees of freedom equal to the number of treatments minus one. In this case, df = 3 - 1 = 2. With $\alpha = .05$ the critical value is chi-square = 5.99.

Friedman Test: As a non-parametric alternative to analysis of variance, the Friedman test uses data from a single sample with each individual participating in three or more treatment conditions. Each individual is ranked from 1 to k across treatments where k is the number of treatment conditions. The following example shows three treatment conditions with n = 5 individuals. Each individual has been ranked 1 to 3 across the three treatments. The sum of the ranks (R) is shown for each treatment.

	Treatments	
I	II	III
3	1	2
3	2	1
2	3	1
3	2	1
3	1	2
R = 14	R = 9	R = 7

Step 1: State the hypotheses and select an alpha level. The hypotheses for the Friedman test do not refer to any specific population parameters.
  H_0: There are no systematic differences among the three treatments.
  H_1: There are consistent differences between the treatments that causes the scores in at least one treatment to be generally higher or lower than the scores in another treatment.
We will use $\alpha = .05$.

Step 2: Locate the critical region. The test statistic for Friedman is distributed as chi-square with degrees of freedom equal to the number of treatments minus one. In this case, df = 3 – 1 = 2. With $\alpha = .05$ the critical value is chi-square = 5.99.

Step 3: Compute the test statistic. The Friedman chi-square is

$$\chi^2 = \frac{12}{nk(k+1)} \Sigma R^2 - 3n(k+1)$$

$$= \frac{12}{5(3)(4)}(14^2 + 9^2 + 7^2) - 3(5)(4)$$

Step 3: Compute the test statistic. The Kruskal-Wallis H is

$$H = \frac{12}{N(N+1)} \left(\Sigma \frac{T^2}{n} \right) - 3(N+1)$$

STATISTICAL TESTS FOR ORDINAL DATA

$$= \frac{12}{15(16)}\left(\frac{36^2}{5} + \frac{22^2}{5} + \frac{62^2}{5}\right) - 3(16)$$

$$= .05(1124.8) - 48$$

$$= 8.24$$

Step 4: Make decision. The obtained value for H (chi-square) is in the critical region. These data are significantly different from chance. Therefore, we reject H_0 and conclude that there are systematic differences among the treatments.

$$= \frac{12}{60}(196 + 81 + 49) - 60$$

$$= 65.20 - 60$$

$$= 5.20$$

Step 4: Make decision. The obtained value for chi-square is not in the critical region. These data are not significantly different from chance. Therefore, we fail to reject H_0 and conclude that there are no systematic differences among the treatments.

HINTS AND CAUTIONS

1. Sometimes it is obvious from looking at the data which sample has the smaller Mann-Whitney U value. Nevertheless, it is wise to compute both U values and check your computations by the formula, $U_A + U_B = n_A n_B$.

2. Remember, to be significant the Mann-Whitney U and the Wilcoxon T must be *equal to or less than* the critical value provided in the table. This is different from most statistical tests where a large value indicates significance.

3. In computing the Wilcoxon T, it is important to remember that the signs should be ignored when ranking the difference scores.

4. Remember that the Kruskal-Wallis test and the Friedman test use a chi-square statistic and that the value for degrees of freedom is determined by the number of treatments, not the number of participants.

SELF TEST

True/False Questions

1. One reason for transforming numerical scores into ranks is that the transformation tends to minimize the negative effects of large variance.

2. A large value for the Mann-Whitney U is an indication of a significant difference between the two groups being compared.

3. The value for the Mann-Whitney U will always be greater than or equal to zero.

4. One way to check you arithmetic when conducting a Mann-Whitney test is to compute a U for both samples (A and B) and verify that $U_A + U_B = n_A n_B$.

5. With large samples it is possible to use the normal distribution to determine the critical region for either a Mann-Whitney or a Wilcoxon test.

6. When ranking the difference scores for a Wilcoxon test, a difference of -4 is considered smaller than a difference of $+2$.

7. When ranking the difference scores for a Wilcoxon test, differences of -1 and $+1$ should receive exactly the same rank.

8. When determining the critical value for the Friedman test, the degrees of freedom are determined by the number of treatments minus one.

9. For a Friedman test comparing three treatment conditions with a sample of $n = 10$ participants, the first step is to rank order the complete set of $N = 30$ scores from 1 to 30.

10. For a Friedman test with $n = 10$ participants, the scores each treatment would be assigned ranks of 1 through 10.

Multiple-Choice Questions

1. The following scores were measured on an interval scale. If the scores are converted to an ordinal scale (ranks), what values should be assigned to the two individuals who started with scores of $X = 5$? Scores: 2, 2, 3, 5, 5, 7, 10, 18
    a. one receives a rank of 4 and the other gets a rank of 5
    b. both should receive a rank of 3.5
    c. both should receive a rank of 4.5
    d. both should receive a rank of 4

2. After obtaining U values for each of the two samples, how is the final U value computed for the Mann-Whitney test?
    a. the larger of the two values: U_A or U_B
    b. the smaller of the two values: U_A or U_B
    c. the product of the two values: $(U_A)(U_B)$
    d. the sum of the two values: $U_A + U_B$

3. An independent-measures experiment with n = 5 in each of two groups produces a value of U = 3 for one group. What is the value of U for the other group?
    a. 5 – 3 = 2
    b. 5x5 – 3 = 22
    c. 5 + 3 = 8
    d. 5x5 + 3 = 28

4. Under what circumstances is the Wilcoxon test used?
    a. Comparing two treatments with an independent-measures design.
    b. Comparing two treatments with a repeated-measures design.
    c. Comparing more than two treatments with an independent-measures design.
    d. Comparing more than two treatments with a repeated-measures design.

5. A Wilcoxon test with a sample of n = 8 participants produces the following set of difference scores: –3, –5, 11, 4, 6, 8, 6, and 9. For these data, what is the value for the Wilcoxon T?
    a. 2
    b. 4
    c. 8
    d. 32

6. What value will be obtained for the Wilcoxon T when the signs of the difference scores are intermixed evenly among ranks?
    a. It will equal zero.
    b. It will be small, near zero.
    c. It will be large.
    d. The value cannot be computed unless some ranks are discarded.

7. What value will be obtained for the Wilcoxon T when the all of the difference scores are positive?
    a. It will equal zero.
    b. It will be small, near zero.
    c. It will be large.
    d. Cannot be determined without additional information.

8. For the Wilcoxon test, what values for T provide convincing evidence of a significant difference between treatments?
    a. Small values near zero
    b. The larger the value, the more likely the difference is significant
    c. Any value that is far from zero, either positive or negative
    d. Any value that is greater than the number listed in the Wilcoxon table

9. The Kruskal-Wallis test is a nonparametric alternative for _____.
    a. the repeated-measures t test
    b. the repeated-measures ANOVA
    c. the single-sample t test
    d. the independent-measures ANOVA

10. For the Kruskal-Wallis test, what values for H provide convincing evidence of a significant difference between treatments.
    a. Small values near zero
    b. The larger the value, the more likely the difference is significant
    c. Any value that is far from zero, either positive or negative
    d. Any value that is greater than the number listed in the Kruskal-Wallis table

11. A researcher obtains numerical scores for three different treatment conditions with n = 8 scores in each treatment. What is the first step if the researcher wants to convert the scores to ranks for a Kruskal-Wallis test?
    a. Combine all the scores into one group and rank them from 1 to 24.
    b. Rank the scores within each treatment from 1 to 8.
    c. Rank the scores across treatments, from 1 to 3, for each participant.

12. A researcher is using a Friedman test to evaluate the difference among three treatment conditions with a sample of n = 15 participants. What will be the value for degrees of freedom for the test statistic?
    a. 2
    b. 14
    c. 28
    d. 42

13. For a Friedman test comparing k = 3 treatments with a sample of n = 10 participants, what is the largest rank that can appear in the data?
    a. 3
    b. 10
    c. 30
    d. cannot be determined without additional information

14. A researcher obtains numerical scores for three different treatment conditions with n = 8 scores in each treatment. What is the first step if the researcher wants to convert the scores to ranks for a Friedman test?
    a. Combine all the scores into one group and rank then from 1 to 24.
    b. Rank the scores within each treatment from 1 to 8.
    c. Rank the scores across treatments, from 1 to 3, for each participant.

15. Which distribution is used to locate the critical value(s) for the Kruskal-Wallis and the Friedman test?
    a. the unit normal distribution
    b. the F distribution
    c. the chi-square distribution
    d. the Studentized range distribution

Other Questions

1. When ranking existing scores, it often is necessary to find ranks for tied scores. Demonstrate this process by assigning ranks to the following scores.
  Scores: 3, 8, 8, 10, 14, 14, 14, 27, 35, 100

2. In the following data, two samples (A and B) have been combined and rank-ordered. Find the U value for each of the two samples.

Rank	Sample
1	A
2	A
3	B
4	A
5	B
6	A
7	A
8	B
9	B
10	A

3. A psychologist performs a study to assess the effect of meaning on memory. One sample of subjects is asked to study a list of nonsense syllables (such as LIF) that have no meaning. A second group studies a similar list and these subjects are told to try to remember the syllables by thinking of something to give it meaning (imagine a person laughing for LAF). Later in the experiment, subjects are given a memory test. The psychologist records the number of items recalled by each subject. For the following data, convert the memory scores to ranks and use a Mann-Whitney test to determine if the treatments are significantly different. Use $\alpha = .05$.

|Sample A|Sample B|
Low Meaning	High Meaning
14	25
6	23
9	19
27	29
16	30
22	24
7	21

4. Rank the following difference scores and find the Wilcoxon T for this sample.
   Difference Scores: -8, 2, -10, -4, 7, -13, -17, -9

5. Briefly explain the two methods for handling difference scores of zero for the Wilcoxon T.

6. A psychologist would like to assess the effect of peer group support on weight reduction. A group of 10 people meets weekly to discuss weight loss and provide emotional and motivational support for one another. After six weeks, the psychologist records their weights to get a preliminary assessment of the effectiveness of the program. Use the Wilcoxon test to determine if a significant change has occurred after just six weeks using $\alpha = .05$. The weights before the experiment and after six weeks of meetings are as follows:

Subject	Before	After
A	201	195
B	158	150
C	151	141
D	150	143
E	171	172
F	146	149
G	162	171
H	147	142
I	145	134
J	150	138

7. A sample of 12 college athletes was selected to compare three different training methods. Five athletes were assigned to each of the three methods. After 2 months the athletes were evaluated and ranked in terms of overall physical fitness. The rankings are as follows:
   Method A: 1, 2, 4, 7
   Method B: 3, 6, 9, 11
   Method C: 5, 8, 10, 12
Use a Kruskal-Wallis test to determine whether these data indicate any significant differences among the three training methods. Use the .05 level of significance.

ANSWERS TO SELF TEST

True/False Answers

1. True
2. False A small value, near zero, is an indication of a real difference between the two samples.
3. True
4. True
5. True
6. False. Difference scores are ranked by magnitude independent of sign.
7. True
8. True
9. False. Rank the three treatments, 1 to 3, for each subject.
10. False. Rank the treatments, 1 to k, for each participant.

Multiple-Choice Answers

1. c Tied scores all receive the average of the tied ranks.
2. b A small value for U indicates a significant difference.
3. b $U_A + U_B = n_A n_B$
4. b Wilcoxon is an alternative to the repeated-measures t test.
5. b The negative differences have ranks of 1 and 3.
6. c If there is no consistent difference between treatments, the Wilcoxon T will be large.
7. a The negative differences will have a sum of ranks equal to zero.
8. a A consistent difference between treatments will produce a T value near zero.
9. d Kruskal-Wallis evaluates the difference between three or more treatments for studies using separate samples
10. b The Kruskal-Wallis H statistic is equivalent to a chi-square statistic. Large values indicate significance.
11. a The Kruskal-Wallis tests ranks the entire set of scores and then substitutes the ranks for the scores in each treatment.
12. a For the Friedman test, df equals the number of treatments minus one.
13. a For the Friedman test, each participant's scores are ranked 1 to k where k is the number of treatment conditions.
14. c Friedman ranks the treatments for each participant.
15. c Kruskal-Wallis and Friedman values are distributed like chi-square with df equal to the number of categories minus one.

Other Answers

1. Scores: 3   8   8   10   14   14   14   27   35   100
   Ranks: 1 2.5 2.5 4   6   6   6   8   9   10

2. The calculation of the points for sample B is as follows:

Rank	Sample	Points for B	
1	A		
2	A		
3	B	------------- 4	Thus, sample B has a total of 9 points.
4	A		Sample A has 15 points.
5	B	------------- 3	
6	A		
7	A		
8	B	------------- 1	
9	B	------------- 1	
10	A		

3. For sample A, U = 7, and for sample B, U = 42. The Mann-Whitney U is 7. With $n_A = n_B = 7$, the two-tailed critical value is 8. Reject the null hypothesis and conclude that meaningfulness does affect memory.

4. When the difference scores are ranked in order of magnitude, it is clear that the positive values have the smaller sum of ranks.

Difference	Rank	
2	1	
–4	2	The positive differences
7	3	have ranks of 1 and 3,
–8	4	so T = 1 + 3 = 4
–9	5	
–10	6	
–13	7	
–17	8	

5. First, you can discard the zeros and reduce the sample size. This method ignores the fact that zero differences tend to support the null hypothesis. Because it throws out data that support H_0, the practice of discarding zeros can increase the chances that you will reject the null hypothesis and increase the risk of a Type I error. The second method involves dividing the zero differences equally between the positive and negative values and ranking the zeros along with the other scores.

6. The null hypothesis states that the weight reduction program has no systematic effect on weight. For these data the positive differences have ranks of 1, 2, and 7. The Wilcoxon T = 10. This value is not in the critical region for n = 10, so we fail to reject the null hypothesis.

7. The T values are 14, 29, and 35 and the chi-square statistic is 4.5. With df = 5 and a critical value of 5.99, fail to reject H_0 and conclude that there are no significant differences.